Cyber CISO Marksmanship

Cyber CISO Marksmanship is the only book of its kind authored by multiple highly experienced individuals to collectively bring together the "best of the best" on what works and what doesn't, as a CISO, in a unique storytelling format. This book is designed for a Chief Information Security Officer (CISO) or an individual seeking this role and also has value to other types of cyber leaders. Knowledge and understanding of traditional books can only get you so far – *Cyber CISO Marksmanship* has powerful perspectives, real-world accounts, and insights you won't find anywhere else!

Key features included with *Cyber CISO Marksmanship:*

- Over a century of CISO experience is represented by the authors of this book
- Unique storytelling format based upon real-world accounts from leading CISOs
- Sharpshooter perspectives from multiple CISOs for each story
- Bottom Line Up Front (BLUF) for quick reference on outcomes for each story
- Sharpshooter icon for what works
- Misfire icon for pitfalls to avoid
- All book owners are invited to participate in online content at **CyberCISOMarksmanship.com** and face-to-face events
- Book owners who are CISOs qualify to join, for free, a private CISO online community (by CISOs for CISOs)

While this book is written for CISOs or those desiring to be in that role soon, it is also helpful to other cyber leaders.

Cyber CISO
Marksmanship
Hitting the Mark in
Cybersecurity Leadership

Ken Dunham, James Johnson,
Joseph McComb, and Jason Elrod

Editorial Expertise by Mark DeBry

CRC Press
Taylor & Francis Group
Boca Raton London New York

CRC Press is an imprint of the
Taylor & Francis Group, an **informa** business

Designed cover image: © Shutterstock

First edition published 2025
by CRC Press
2385 NW Executive Center Drive, Suite 320, Boca Raton FL 33431

and by CRC Press
4 Park Square, Milton Park, Abingdon, Oxon, OX14 4RN

CRC Press is an imprint of Taylor & Francis Group, LLC

Library of Congress Cataloging-in-Publication Data
Library of Congress Cataloging-in-Publication Data
Names: Dunham, Ken, author. | Johnson, James (Chief Information Officer),
author. | McComb, Joseph, author. | Elrod, Jason, author.
Title: Cyber CISO marksmanship : hitting the mark in cybersecurity
leadership / Ken Dunham, James Johnson, Joseph McComb and Jason Elrod.
Other titles: Cyber chief security information officer
Description: First edition. | Boca Raton : CRC Press, 2025. | Includes
bibliographical references and index.
Identifiers: LCCN 2024029355 (print) | LCCN 2024029356 (ebook) |
ISBN 9781032720425 (hardback) | ISBN 9781032720463 (paperback) |
ISBN 9781032720500 (ebook)
Subjects: LCSH: Computer security--Management. | Computer
networks--Management.
Classification: LCC QA76.9.A25 D848 2025 (print) | LCC QA76.9.A25 (ebook) |
DDC 658.4/78--dc23/eng/20240821
LC record available at https://lccn.loc.gov/2024029355
LC ebook record available at https://lccn.loc.gov/2024029356

ISBN: 978-1-032-72042-5 (hbk)
ISBN: 978-1-032-72046-3 (pbk)
ISBN: 978-1-032-72050-0 (ebk)

DOI: 10.1201/9781032720500

Typeset in Sabon
by KnowledgeWorks Global Ltd.

"Jesus looked at them and said, 'With man this is impossible, but with God all things are possible'" NIV Matthew 19:26

I'd also like to thank all the amazing people who have invested in me and helped me along the way over all these years. What a journey it has been! It would not have been nearly as fun without you, nor was it possible – I treasure each one of you! I would especially like to thank my family, calling out my mother and father (Jack and Susan Dunham), who have invested and believed in me the most, teaching me to persevere and overcome – especially in the most difficult of times – and serve others in all you do.

Ken Dunham

To my grandchildren Waylon, Lacey, Kaylee, and Colt who remind me daily the importance of documenting successes and failures so they can build upon them and continue to make the world a better place for all.

James Johnson

To my mom and dad: Dad, thanks for showing me the world. Mom, you made my day when you asked me to freeze your credit.

Joseph McComb

To all who are struggling, whether known or unknown, spoken or unspoken: You're not alone. Remember, it doesn't matter how many times you fall down, only how many times you get up.

I'd also like to thank my mom for being the original superhero in my life. Thank you for showing me how to make the best of every situation and to not waste time on the things that don't deserve it.

Jason Elrod

Contents

Foreword

"A true leader has the confidence to stand alone, the courage to make tough decisions, and the compassion to listen to the needs of others. He (or she) does not set out to be a leader, but becomes one by the equality of his (or her) actions and the integrity of his (or her) intent."

Douglas MacArthur

The only place where success comes before work is in the dictionary. In *Cyber CISO Marksmanship*, the authors share from their hard-fought battles the lessons they have learned over decades of time on the Cyber battlefield. Their work, which they express so thoughtfully in this book, is your path to accelerate the success you can have as a CISO.

I have long believed there are two battlefields a CISO faces daily. The external battlefield of threat actors and vulnerabilities that we see every second of every day whether it is from a threat intelligence feed, an alert from the array of cybersecurity tools we have deployed, or from a risk assessment the security team performed. On this battlefield, your entire organization is engaged every second of every day. It is one where we can become weary, but we realize our strength comes from the struggle.

But there is another battlefield. It is the internal battlefield of budgets, bureaucracy, and behavior. This battlefield is lonelier. While most of the security team has their aim focused on the external battlefield, it is the marksmanship of the CISO on the internal battlefield that can, in most cases, decide the fate of risk for the organizations we support. It can also determine the fate of the CISO themselves when it comes to operating with integrity.

This is not just another security book. Stories are data with a soul. Ken, James, Joe, and Jason share their stories on both battlefields so that we can learn, so our work can be more effective, and ultimately so we can succeed. They are true leaders.

Malcom Harkins
"Sharpshooter" expert,
Chief Security and Trust Officer,
HiddenLayer.

Preface

Ken Dunham, one of the authors of this book, while working as a consultant with F100 organizations regularly performed vCISO services for organizations. In this capacity, Mr. Dunham and respected colleagues discussed gaps regularly seen in leadership, where the industry commonly missed the mark, and where leaders were commonly struggling to find support. Mr. Dunham brought together industry leaders to collaborate and establish a global community to fill this gap, so that global CISOs would no longer be alone while striving to create change in a complex and challenging role.

Beyond the standard technical and regulatory expertise, you would expect from a book of this nature, the co-authors, Ken Dunham, James Johnson, Joseph McComb, and Jason Elrod, harness the power of narrative and metaphor to inspire lasting change and drive meaningful action in the realm of executive leadership. This book offers not just advice but mentorship, drawing on the extensive experience of the co-authors to provide readers with a roadmap to leadership excellence. The authors, having climbed the ranks from the ground up, possess intimate knowledge of the challenges and triumphs at each stage of a cybersecurity professional's career. This perspective makes them uniquely equipped to guide leaders at all levels, from those taking their first steps into management to seasoned executives.

Storytelling is another unique angle used in the approach for this book. In over 30 years of leadership and consultation, Mr. Dunham has experienced storytelling as an essential part of personal relationships and connections. This powerful form of messaging has manifested itself in *Cyber CISO Marksmanship* with the wisdom and insight of four different CISO personalities and backgrounds. Within the book, our CISO Sharpshooters provide a wealth of valued inputs and information that every CISO will treasure.

This book should not be used as a substitute for the advice of a competent attorney admitted to or authorized to practice in your jurisdiction. Stories are based on true stories and experiences by authors and supporting contributors. Stories may include a compilation of collective author experiences and interpretations that are fictitious to a given individual and not relevant to any real-world experience with any specific author or contributor to this book. Stories may vary from real-world events any given individual may have experienced with authors. Authors have chosen a storytelling format to connect deeply with their audience, to tell the truth as it needs to be revealed, where it matters most.

About the authors

Ken Dunham

Ken Dunham, CEO of 4D5A Security, has over 30 years of global cyber leadership experience, including executive leadership in a leading America's security company, key involvement in two top-rated startups, innovation of responsible disclosure used by Microsoft and others for vulnerability management today, innovation of cyber threat intelligence (CTI), and extensive incident response and counterintelligence experience within F100 and other organizations around the world. Mr. Dunham also innovated groundbreaking training programs integrating new forms of technology and training for the USAF for the U2 spy plane, Warthog, and creation of the Predator drone program. Mr. Dunham is the author of seven books, a top-rated website, and the top downloaded antivirus program. He is a recognized global leader within ISSA as an International Distinguished Fellow, authoring multiple books and the ISSA international article of the year, "Troubling Trends of Espionage," 2015.
https://www.linkedin.com/in/4d5asecurity/

Jason Elrod

Jason Elrod is the Founder and CEO of Hanging Rose Research, an independent think tank and advisory firm focused on executive leadership and cybersecurity innovation. He has over 30 years of experience in business and cybersecurity leadership across multiple sectors, including healthcare, finance, telecommunications, and information technology services. Mr. Elrod has been a board member, board advisor, lead security executive, and CISO for multiple organizations and institutions. He holds advanced certifications and degrees in information technology and cybersecurity, and he is a frequent speaker and participant in industry conferences and conversations.
https://linkedin.com/in/thejasonelrod/

James Johnson

James Johnson is the CIO and previous CISO of Holland & Hart, the largest legal firm headquartered in the Mountain West. He has 25 years of IT and cyber executive experience in Fortune 500 companies, where he has led organizations in the government and private sectors. He spent 3 years as a cyber security consultant working with e-commerce companies, cities, universities, and airports to improve IT and security processes. Mr. Johnson developed and implemented a Security Strategic Plan for Denver International Airport. During his tenure as President and Chairman of the Denver ISSA Board of Directors, the chapter grew to the world's largest and received the Chapter of the Year Award. He continues to serve ISSA as the Co-Chair of the Rocky Mountain Information Security Conference (RMISC).

https://www.linkedin.com/in/jamesjohnsonpcss/

Joseph McComb

Joe McComb, Ph.D., is the Chief Information Security Officer at Holland & Hart LLP, a position he has served in since August 2021. McComb oversees the firm's cybersecurity, data privacy, and awareness programs and management. Before joining Holland & Hart, McComb served as the CISO of Ball Aerospace and the Global CISO for Janus Henderson Investors. McComb holds a bachelor's degree, summa cum laude, in biochemistry from the University of Colorado and a master's and Ph.D. in physical and biological anthropology from the University of Kansas. Over the last 30 years, Joe has released diverse publications, including Education Theory, Anthropology, and Cybersecurity. He holds multiple certifications, including the GSEC, G2700, CISM, CISSP, and the CISA.

https://www.linkedin.com/in/joe-mccomb-80667a/

Special acknowledgments

EDITORIAL EXPERTISE

Mark DeBry

Mark DeBry is the VP of Business Operations, vCISO, and Certified CMMC Assessor at Shadowscape, an intelligence-driven cybersecurity services, training, and analytics company. He is an experienced global IT, cloud, and cybersecurity executive with many years building enterprise-wide programs at IBM and Microsoft. He has led the strategy and implementation of IT and cybersecurity frameworks, governance, risk, compliance, and technical controls. Mark has built and managed sales, solutions, and delivery teams in the US, India, Argentina, Costa Rica, and Brazil. He has several certifications, including the CISM, PMP, as well as Microsoft Azure and AWS Cloud Certified. Mark also has a master's degree in information systems management.
https://www.linkedin.com/in/markwdebry/

SHARPSHOOTERS

Brad Bussie

Brad Bussie is an award-winning 20-year veteran in the information security industry. He is an author, Chief Information Security Officer, and industry thought leader.

He holds an undergraduate degree in information systems security and an MBA in technology management. Brad possesses premier certifications from multiple vendors, including the CISSP from ISC2. He has a deep background in cybersecurity, identity/access management, vulnerability management, governance, risk, and compliance. Brad has spoken at industry events around the globe and has helped commercial, federal, intelligence, and DoD leaders solve complex security challenges.

His mission is to provide security leaders with practical advice on how to protect their organizations from cyber threats more effectively. Brad makes understanding security as simple as possible without making it simpler.
https://linkedin.com/in/bradbussie/

Shawn Murray

Shawn Murray is President and CAO at Murray Security Services and was previously assigned to the United States Missile Defense Agency as a senior cybersecurity professional. His previous assignments included work with the US Army Cyber Command in Europe, the US Air Force, and the commercial industry in various information assurance and cybersecurity roles. He has traveled the globe performing physical and cybersecurity assessments on critical national defense and coalition programs, has prepared reports for the House Armed Services Committee, and has testified for the US Senate on small business concerns. He is the lead cyber consultant at the Pikes Peak Small Business Development Center (SBDC) and sits on the national SBDC cyber working group.

Dr. Murray has worked with SBA, the Colorado Attorney General's office, the NSA, FBI, CIA, and the US Defense and State Departments on various Cyber initiatives and has over 20 years of IT, communications, and cybersecurity experience. He has presented as a featured or keynote speaker for numerous conferences across the globe. He enjoys teaching and presenting as a guest lecturer on cybersecurity, business, and computer science courses at his Cyber Academy and for several universities. He has several industry-recognized certifications, including the C|CISO, CISSP, and CRISC. He holds several degrees including an Applied Doctorate in Computer Science with a concentration in Enterprise Information Systems.

Dr. Murray is a Distinguished Fellow at the Information Systems Security Association and was recently elected as the new President of the International Board of Directors. He is a member of ISC2, ACM, IEEE, ISACA, and the Colorado Springs World Affairs Council. He enjoys spending time traveling with his family, researching and collaborating with other professionals in cybersecurity and Cyber Law, and playing soccer in a local league in Colorado Springs.

https://www.linkedin.com/in/dr-shawn-p-murray-3a618a13/

Malcolm Harkins

Malcolm Harkins is the Chief Security and Trust Officer at HiddenLayer. In this role, he reports to the CEO and is responsible for enabling business growth through trusted infrastructure, systems, and business processes. Malcolm is also responsible for peer outreach activities to drive global improvement in understanding cyber risks best practices to manage and mitigate those risks. He is also an independent board member, advisor to several organizations, and CISO Ambassador for Reveald. He enjoys being an executive coach to CISOs and others in a wide variety of information risk roles. Key areas of focus include the ethics around technology risk, social responsibility, total cost of controls, public policy, and driving more industry accountability.

Previously Malcolm was the Chief Security and Trust Officer at Cylance. Malcolm was previously Vice President and Chief Security and Privacy Officer (CSPO) at Intel Corporation. In that role, Malcolm was responsible for managing the risk, controls, privacy, security, and other related compliance activities for all of Intel's information assets, products, and services.

Before becoming Intel's first CSPO, he was the Chief Information Security Officer (CISO). Over Malcolm's 24 years with Intel, he also held roles in finance, procurement, and various business operations. He managed IT benchmarking initiatives and Sarbanes-Oxley compliance; was the profit and loss manager for the Flash Product Group; was the general manager of Enterprise Capabilities, responsible for the delivery and support of Intel's Finance and HR systems; and he helped create an Intel business venture focusing on e-commerce hosting in the late 1990s.

Malcolm previously taught at the CIO Institute at the UCLA Anderson School of Management and was an adjunct faculty member at Susquehanna University in 2009. In 2010, he received the RSA Conference Excellence in the Field of Security Practices Award. He was recognized by *Computerworld* as one of the Premier 100 Information Technology Leaders for 2012. (ISC)[2] recognized Malcolm in 2012 with the Information Security Leadership Award. In September 2013, Malcolm was recognized as one of the Top 10 Breakaway Leaders at the Global CISO Executive Summit. In November 2015, he received the Security Advisor Alliance Excellence in Innovation Award. In 2023, he received the Cyber Defense Magazine award for the top Chief Security Officer. He is a Fellow of the Institute for Critical Infrastructure Technology, a nonpartisan think-tank providing cybersecurity to the House, Senate, and a variety of federal agencies. Malcolm is a sought-after speaker for industry events. He has authored many white papers and, in December 2012, published his first book, *Managing Risk and Information Security: Protect to Enable®*. He was also a contributing author to *Introduction to IT Privacy*, published in 2014 by the International Association of Privacy Professionals. The 2nd edition of Malcolm's book, *Managing Risk and Information Security: Protect to Enable®*, was published in August of 2016. Malcolm has testified before the United States Senate Committee on Commerce, Science, and Transportation on the "Promises and Perils of Emerging Technology for Cybersecurity." He has also testified at the Federal Trade Commission hearings on data security. In 2023, Malcolm was a member of a task force led by the Center for Strategic International Studies to provide strategic direction and leadership for CISA's evolving mission to protect the federal government.

Malcolm is on the board of directors for TrustMAPP, Cyvatar, and the Cyber Risk Alliance.

Malcolm received his bachelor's degree in economics from the University of California at Irvine and an MBA in finance and accounting from the University of California at Davis.

https://www.linkedin.com/in/malcolmharkins/

Acknowledgments

The authors of the book would like to thank the innumerable number of individuals who have supported us through the years enabling us to be successful throughout our career. This includes both those that we enjoyed working with and those that were difficult, pushing us to learn in different ways, and improving our character. We give thanks to all of you, for helping shape and mold us into the leaders that we are today, to be the next best person we can be today to achieve and contribute toward team outcomes and business success.

Cyber CISO Marksmanship would not have been possible without the support of the following:

Sharpshooters:

- Brad Bussie, CISO, e360
- Tim Coogan, CISO, Colorado Regional Transportation District (RTD)
- Malcom Harkins, Chief Security & Trust Officer, HiddenLayer
- Dr. Shawn P. Murray, President, International, Information Systems Security Association (ISSA)

Family, Friends, and Those Supportive of the Authors:

- From Ken Dunham: My amazing children – now all young adults – Michelle, Ben, Josh, Noah, and Amberly. Being your Father taught me lessons of love, humility, grace, and surrender that can only be learned through the joys, failures, and success of parenting and family. Thank you for every moment, good and bad, that shaped us into who we are today. Love Dad.
- From James Johnson: To my much better half, Mary Ann, who has stood by me through the years as I worked crazy hours, nights, and weekends as required by the CISO and other IT positions. To my first boss and father figure, Clifton Bell, who took a really young, poor kid and taught him the value of listening to the customer, the satisfaction of doing a job to the best of his ability, and the value of hard work.

- From Joseph McComb: My wife, Natalie, who puts up with me constantly typing on a keyboard at all hours of the night.
- From Jason Elrod: A heartfelt thanks to my amazing wife Debi, my family, and my friends, who are the eternal champions of my offbeat inspirations (and sometimes downright crazy ideas). This book wouldn't exist without your encouragement for me to chase them all.

Introduction

From one Chief Information Security Officer (CISO) to another, welcome to the book and community you've been looking for to help you understand "what works" in a CISO's world. Unlike other books that focus on academic theories and a catalog of threats and malware, *Cyber CISO Marksmanship* offers something unique. Highly experienced CISOs share their journeys through captivating real-world stories, providing invaluable insights and practical advice. Dive in and discover the strategies that have proven successful in the dynamic and challenging realm of cybersecurity leadership.

Imagine meeting one of the authors at a conference and sitting down for a drink or meal to discuss a cybersecurity topic. Diving in further, you would work to develop a trusted relationship, connecting on your leadership background, personalities, areas of mutual leadership, and "what works." That's precisely what this book is about. Real-world stories about what really happened in the war room, the incident you never heard about, the strange interview, or the heated boardroom exchange.

These experiences can help you learn to overcome challenges, gain confidence, and feel validated in your experiences. As a CISO, you'll realize you belong. You'll build confidence in the right areas and dispel the beliefs that hold you back. You'll learn to plan, prioritize, and lead effectively, moving to the left of the boom. You'll transform from being merely busy to being influential and leading. Your teams will shift from being disparate to unified, working together with a mission toward both short- and long-term goals.

Being a CISO is not for the faint of heart. We invite owners of this book, who have the proper roles and responsibilities, to join our exclusive "By CISOs for CISOs" community. In this community, you'll collaborate with other invited CISOs. We plan to meet face-to-face around major events and conferences to support and build each other up organically. You are not alone. It takes a village to build a SecOps program. You no longer need to feel isolated or uncertain about whom to trust for help. Your global Cyber CISO Marksmanship network, composed of highly experienced CISOs, is here to help you always hit the bullseye!

This book and its stories are invaluable resources for anyone passionate about cybersecurity, especially those aspiring to become CISOs. These

DOI: 10.1201/9781032720500-1

seasoned CISOs have navigated priorities, battled adversaries, and emerged stronger from the refiner's fire. Their experiences offer profound lessons and practical insights, helping you avoid many of the mistakes they've encountered. Join in and learn from the best to shape your path toward becoming a more effective cybersecurity leader.

JOIN OUR ONLINE COMMUNITY:
CyberCISOMarksmanship.com

If you're a CISO and own this book, you are qualified to join our exclusive online community, "By CISOs for CISOs." In this private community, we collaborate, network, and support each other globally through a dedicated mailing list, helping one another overcome challenges and share insights.

There is also a mailing list exclusive to current CISOs. The authors have also pre-recorded FIRESIDE chats, like grabbing a drink with colleagues at a conference, to discuss each chapter of the book. These chats are available online to those who join the community. Additionally, the authors will host regular online events to foster ongoing collaboration and community for CISOs.

We hope you'll join us!

Chapter 1

Defining the CISO role

WHAT CONSTITUTES A CISO?

The industry generally recognizes the Chief Information Security Officer (CISO) as the organization's executive leader of information security. The CISO must align security strategies to help drive business objectives while managing organizational risks. As companies have different business strategies and associated risks, CISO duties may vary dramatically from company to company. Some CISOs have physical security and business continuity in their area, while others mainly cover IT, policy, and governance. Some positions require a CISO with strong technical skills, others require executive management experience, while some need deep regulatory knowledge. Since the role requires significant expertise covering a broad and expanding field, CISOs must possess an insatiable desire to learn. This chapter explores many areas defining a CISO, including job duties, certifications, technology, management, and executive presence.

⮂ First day as CISO

I was promoted and moved to headquarters for the newly created CISO position. Upon arrival, there was no job description, and my boss couldn't explain the expectations of the position. I found out they hadn't even determined where I would report. I expected the VP of IT to be my new manager, but he decided that the CISO position would report to the Governance Director. With that settled, I headed to my first meeting for an introduction to the field engineers who were responsible for implementing IT solutions at our field sites. It turned out to be more of a complaint session where I listened to all their concerns about how security would negatively impact their work. Once they finished, I repeated their concerns and thanked them for being honest. I emphasized how valuable their input was to the program and reassured them that they would have input as stakeholders. I felt I'd handled that situation well and headed to my office to call the only other person working on security. On the way, the Governance Director called and told me that I needed to sign off on a project immediately so they could get started. I asked a few probing questions and discovered the project was already underway. Through this discussion, I quickly learned that the CISO was expected to accept all IT security risks, whether or not they were involved in the project planning and design. To make matters worse, I learned my approval was to shield the VP of IT if something went wrong. That day, I also learned that the parent company had recently emerged from bankruptcy, so there was significant focus on being traded again on the New York Stock Exchange (NYSE). The sole focus became profitability to increase the stock price rapidly. I would later determine during the first few months that support from upper management for security could be summarized as, "Sure, I support security. As long as it takes no budget, no staff, and does not create any project delays." I'd made the classic rookie mistake of not researching this critical information upfront. I thought, "What did I get myself into?" and "How can I handle the situation to increase my odds of success?"

Sharpshooter perspectives (SSP)

SSP1: This story reminds me of my red flags, which indicate that I might be working for a company seeking a scapegoat when things go awry. Here are the red flags I keep an eye out for:

- Lack of authority and resources
- Absence of clear communication and reporting lines
- Minimal involvement in strategic decision-making
- No support during security incidents

SSP2: I remember a breach caused by a structured query language (SQL) injection attack. The CTO) asked, "How can we prevent this in the future?" I laid out a simple diagram of secure development processes, including testing steps. To his credit, the CTO had me present it to the developers and reinforced that this is how we need to do things now. Unfortunately, that CTO left a month later, and his successor didn't care. Genuine management support is crucial.

SSP3: As a CISO, I've heard someone above me say, "No problem. If we get breached, we'll just fire the CISO." This abrasive attitude sends a message that some people in the organization don't care about security and don't have your back.

SSP4: I've experienced situations where the responsible person above me refuses to sign off on a risk. I've also had senior leadership ignore requests during application development to require encryption of credit card and social security numbers. They took this approach, hoping they could evade responsibility for the situation if things went wrong. To protect myself, I always document risks so there is always a record if someone tries to pin an incident on me.

Target grouping

During my tenure at a previous company, I led the development of a highly successful cybersecurity program. This led to my continued employment with the acquiring company for 5 years, during which their stock price increased tenfold. The acquisition was partly motivated by our established processes, including cybersecurity, with the program I implemented becoming the foundation of the new company's cybersecurity strategy.

On a call with my one direct report, I inquired how risk was managed for the company, and he told me that the Corporate Risk Committee managed all the organization's risks except cybersecurity. At the same time, the VP of IT was responsible. I explained that the Governance Director made it clear that I was responsible for all computer security risks, or in other words, being responsible for all the risks they introduced into the environment, so it was apparent I had been hired so the VP of IT could transfer the risk to me. He wasn't surprised and told me he was committed to maturing the program and ensuring a proper risk management and acceptance process. He also informed me that the field engineers are under tremendous pressure to build remote sites in days, and they wanted to avoid any barriers that security might cause and delay the timeline.

To bring some sanity to cybersecurity risk, I developed a risk program patterned after the company's construction risk program so the executives and the Corporate Risk Committee could easily understand it. The VP of IT wasn't interested in the risk program and took the approach that he could transfer the risk with insurance. He set up a meeting to include our insurance carrier, a representative of the Corporate Risk Committee, himself, and me. The insurance company explained that a risk mitigation program was required before they would insure an event and praised the program I presented as a model for all their clients to follow. The Corporate Risk Committee made implementing my program a requirement and held the VP of IT responsible. The risk program was the genesis for developing a comprehensive governance model to mature the cybersecurity program.

Despite all the challenges, I remained enthusiastic about the position and worked tirelessly to develop and mature the program, understanding that security is a journey. I implemented the security and management principles I had been taught and utilized the helpful advice of my peers. Even though we built a successful security program, I wish I'd had this book and applied many of the concepts in this book on that fateful first day as CISO.

⊕ Sharpshooter Tips

- No matter what situation you encounter as a CISO, maintain professionalism, structure your cybersecurity program using the methods and controls required to protect the organization, continuously drive the program forward, and never compromise your integrity.

- Understanding the compliance frameworks that drive the control requirements will allow you to align your governance program with the company's regulations.
- As a new CISO or head of cybersecurity, clearly understand where you will report in the organization. The higher you report, the more influence and impact you will have.

⌐ Misfires

- Starting a new CISO role without precise requirements and objectives makes developing a program roadmap difficult, leading to unnecessary confusion and inefficiency.
- Failing to thoroughly understand the organization's structure, culture, and security needs can hinder your ability to develop and implement effective security measures.

⊇ Why should we hire you?

On my flight to the on-site interview with a group of executives, I tried to anticipate their interview questions. I researched which individuals would be in the room, their professional expertise, and what they would likely value in soft and hard skills for my leadership role. I met my escort, badged into the facility, and was introduced to the interviewing group. I felt comfortable with the group and was excited about the interview. During the interview, one person asked me, "Tell us more about your education and qualifications for this role. After all, your degree is in education, not computing. Why should we hire you?"

Sharpshooter perspectives

SSP1: This question shouldn't come as a surprise, particularly for interviewers who lack familiarity with cybersecurity. A quick LinkedIn search shows that many CISOs and lead security personnel did not start in a technical role. I would key off the word "qualifications" and go through my list of successes that show "I can do the job because I've already done it before." I'd also mention that technology is rapidly changing, so someone with technical skills from years ago can lose their effectiveness if they aren't keeping up with the latest technology. Explain how you're involved in cybersecurity training for those new to the industry. If that doesn't appear to be getting through, use a little humor about your degree, in that CISOs are like third-grade teachers, tailoring their training materials and message at an 8- and 9-year-old level since most employees know security at this level. While the interviewer focused on technical knowledge and ability, I'd explain how well I connect with people and their need to succeed, my understanding of the business, and my delivery of security solutions that enhance corporate success.

SSP2: I love questions like this. They've just opened the door for you to say whatever you want, which means you should have a powerful "2-minute pitch" ready to roll. Like a politician, pivot and say, "Great question. My academic experience has taught me {fill in whatever cool thing you want} which makes me uniquely qualified to manage the position's responsibilities, unlike those who don't have the wider life and educational diversity that comes along with that." Turn a perceived weakness into a strength when you can.

SSP3: I have gotten similar shades of this question, which tells me much about the organization. Twenty years ago, information security degrees were scarce, which meant many security professionals lacked formal education in the field. Consequently, most security roles paid less than other IT positions. Back then, you went into information security because you enjoyed the work! In general, there are two types of interviewers: (1) the ones who are genuinely interested in finding out if I'm the best person for the role and if I can do the job, and (2) the ones who have already decided I'm not the right person and are working to convince the other interviewers.

For the interviewers interested in finding out if I'm a good fit, I outline how my education set me up with practical skills in interpreting data, managing people, and educating individuals. Those who have already decided I am not fit are usually hostile in their approach. Typically, I use the same outline but observe how they react. On the flip side, I'm looking for red flags and contemplating whether I want to be part of their organization based on how they treat me during the interview.

Target grouping

The interview question had a negative undertone, suggesting I was not qualified for the role. I had an important choice to make during the interview on my soft skills and how I would respond to the question. In my heart, it would be easy to be angry, offended, or even afraid or doubt myself, but I chose to be self-aware, realize that it is a legitimate question, and be thankful that they were transparent enough to ask. I surmised that they were wrestling with hiring me when I wasn't the traditional candidate compared to everyone else.

I responded professionally, thanked them for the question, and then informed them that my experience, combined education and on-the-job training, more than compensated for what any degree would have taught me. Furthermore, my career accomplishments far exceed any degree and any other candidate. I declined the job offer because it didn't align with my skills and career goals compared to the other positions I had applied to.

In the early days of cyber, very few had experience because the field was emerging. We all had to learn by doing. In the "wild-west" days, you could find a job quickly because certifications and degrees didn't exist or matter as much as they do today. Securing a prominent role such as a CISO is feasible without a traditional degree, but it's becoming more challenging as you contend with numerous candidates who have taken the conventional security degree and career path. The best candidate continually learns, demonstrates a passion for expanding their education and certifications, and becomes highly skilled with a strong balance between educational and real-world experiences.

⊕ Sharpshooter Tip

- Respond instead of react, and be the best "you" as a leader. In this context, you are interviewing them as much as they are interviewing you. Think about how lucky they would be to have you as an asset to their team. Show the best you in every response you give as a leader, acting fearlessly and not allowing yourself to be intimidated by others' hostility, intimidation, or doubts. When executing in a group, leadership in the fire has no room for fear.

↝ Misfire

- Career-limiting self-doubts exist in all of us. As a leader, acknowledge your doubts, address them, put them behind you, and move on to tackling challenges. Don't allow your doubts, or more importantly, the doubts of others, to ever cast a shadow on your proven excellence as a leader. Be the light you are as a leader, and let it shine without limits.

⊇ Certification and credibility

Over the years, I've discussed the value of certification with other CISOs. Many CISOs have obtained the industry-recognized Certified Information Systems Security Professional (CISSP). As an information security lead, I took my CISSP on a snowy day in December in the early 2000s. A month later, with CISSP in hand, friends began to send me jobs they felt I should apply for. As a security professional, the CISSP opened many doors over the years.

As an executive, I found it more challenging to find time for certifications. The company where I held my first executive information security role put minimal emphasis on certifications or continuous learning. I was falling behind technically due to the lack of continuing education because training budgets were slashed. I picked up the Certified Information Security Manager (CISM) textbook and forced myself to study by registering for the exam in a few months. Work was busy, and soon, I found myself days before the exam with little study.

Fortunately, 2 days before the exam, I attended an information security committee meeting on the other side of the country and had two 4-hour plane flights. I pulled out my textbook and studied the entire flight there and back. I took the CISM test the next day and passed. Several weeks later, the company board inquired about the information security team's qualifications, and I provided the CISM as one of our certifications.

Sharpshooter perspectives

SSP1: Certifications are essential for the cybersecurity professional, whether on the technology or management path. I have completed numerous security certifications, including GIAC Security Essentials (GSEC), GIAC Security Leadership (GSLC), and the CISSP. Of the three, the CISSP has been viewed as the most credible and has had the most positive influence on my CISO career. I dropped the GSEC and the GSLC since I considered these more entry-level certifications and the ongoing costs of taking additional SANS courses for maintaining them to be ridiculous. Certifications outside of security can also be beneficial to the CISO. When working with an engineering and construction company, I was the only person outside the engineering group who took the company's rigorous project management certification program, which took 2 years of study, multiple written tests, and an oral final exam. That resulted in a lot of company credibility. Still, with that experience, I decided to pursue the Project Management Professional (PMP) for external credibility, and that certification has been invaluable for numerous positions in multiple industries. One of my mentors was a retired US Navy Submarine Commander, and he stated that the PMP was the most challenging exam he ever took, and there was no way he would ever miss the requirements for renewal.

SSP2: If you don't have the CISSP, get it. Beyond that, I don't think any other certification will make a difference to you as a CISO, even if a recruiter requires another certification that helps you get a ticket to the dance (interview). That said, the value of any certification is the knowledge acquired to get it and not the actual certification itself. I've told my employees who ask the company to pay for training and certification that I will pay for their training but not the certification because the training benefits both the organization and the employee, but the certification only benefits the employee. The lesson here is that if they want the certification paid for, then make sure you select a training that includes it.

Target grouping

A CISO must prioritize their time to achieve and maintain certifications, establishing themselves as a leader among cyber leaders. Certifications demonstrate to the board that you are a credible information security professional.

There is much debate on the practicality of certifications and whether a company should pay their employees to become certified. In general, security regulations within the US push for board members to have access to cybersecurity expertise. One way to demonstrate this is through cybersecurity certifications. All company boards I have worked with have asked for my cybersecurity certifications. Unfortunately, not every company has been willing to support continuing education in cybersecurity. As I have delved into why some upper managers don't see continuing education as necessary, I have noticed that they typically expect that I know it already. Oddly enough, the companies I have worked for that have not supported continuing education are also the most behind in technology and security – which says something.

⊕ Sharpshooter Tip

- Continuing education is critical for both you and your staff. Determine the company support for cybersecurity certification and, if required within your industry, push for funding as part of the overall program budget.

⌐～⌐ Misfire

- A potential downside is hiring a well-credentialed individual with little experience. In interviewing, ask for concrete examples to know that the security analyst is up for the task!

THE MULTIPLE HATS OF THE CISO

The CISO role is an executive function that manages security risk while enabling and protecting business strategy. The security roadmap must align with the business strategy and keep the business operational. Although their compliance duties may put the CISO at odds with other executives, ultimately, the CISO is part of the business, driving success and balancing risk. Multiple responsibilities result in a complex role that often requires knowledge of technology, regulations, risk management, finance, physical security, and business continuity. Information security is embedded in multiple areas across the enterprise, making it challenging for any single practitioner to have deep knowledge across all disciplines.

Given the complexity of the role, a CISO must foster a culture of collaboration, relying on others, such as IT, to assist in implementing security controls and with Operations to ensure security doesn't tip the scale and jeopardize revenue-generating activities. Maintaining a solid set of metrics to monitor the program's effectiveness is crucial. Many firms also have privacy regulations, such as HIPAA or GDPR, that necessitate the security program's understanding of protecting a client's personal data and establishing data loss prevention (DLP) to ensure data is not inadvertently compromised. When the volume is overwhelming, CISOs may need to consider additional support or outsourcing work to other organizations, highlighting the need for effective collaboration.

This section includes examples of data privacy and measuring vendor performance. Balancing risk within a company is a complex endeavor, and the CISO has to walk a fine line between implementing controls that meet regulatory compliance and being seen as a hindrance to the business. Later, we'll illustrate the balance between being seen as the "No" department or implementing the right level of security that protects the company while allowing it to grow.

ම A tale of two cultures

When integrating an information security program from Europe with one from the United States, I faced the challenge of merging the respective privacy programs. Because there were no other executives willing to champion it, Privacy ended up under my responsibility. On my first trip to Europe, the US Head of Data Privacy gave me a mission: find out who was in charge of privacy at the company we were merging with.

In mid-2016, the US privacy team completed an analysis of the impact of the General Data Protection Regulation (GDPR) and found that less than 1,000 of our clients fell under the regulation. In January 2017, my company was merging with a European-based firm, and compliance with GDPR loomed in May 2018, meaning it was now a high priority.

In Europe, the information security team had recommended a consulting project to determine control readiness for GDPR. As I worked through the pieces, I compiled the following facts: (1) the person who supported privacy at the other firm was an attorney who maintained the company privacy policy but was not a dedicated resource; (2) the other firm had over a quarter million clients who were in the European Union; (3) there was hostility in having a US team heading information security and privacy. A direct quote was, "What does the US know about privacy?"

Sharpshooter perspectives

SSP1: I've had similar challenges working with UK colleagues who treated US consultants with bias based on their cultural perceptions. In some cases, because of their own local, regional, or EU-wide laws like GDPR, they falsely believe they are privacy experts when, in fact, privacy impacts everyone, including those in the UK and the US. This challenging situation required some tough talks with senior program managers and leaders to overcome bias and hostility up front to ensure a successful outcome. GDPR compliance at scale is exceptionally complicated and challenging. Considering the magnitude, I was extremely concerned about meeting deadlines and staying within budget, especially without a dedicated resource and challenging biases.

SSP2: A US company merging with a European-based company can have significant privacy risk implications, especially in specific verticals such as personal finance and healthcare. This merger can become a considerable business risk since fines can be up to 20 million euros or, in the case of an undertaking, up to 4% of the total global sales the preceding year. Depending upon the size of the business being acquired, subjecting the US entity to this risk may not be worth it and must be considered in the mergers and acquisition details of the deal. A significant presence in Europe will likely require the US business to become GDPR compliant by May 2018, which adds complexity to the compliance plan.

As the leader of the US team, I would take the other team's suggestions for a third-party assessment, even if it was just to establish goodwill. I didn't think either team had the expertise or the resources to fully determine the existing gaps and be able to develop a compliance plan that could be implemented in less than 18 months. Since there was a much greater understanding of GDPR requirements in Europe versus the US, there should be expertise that could be leveraged from the other team to help facilitate compliance in the US. Also, flying European team members to the US would signal their collaboration and hopefully reduce their defensive attitude.

SSP3: Culture wars can lead to more compliance problems than differences in GDPR and US privacy laws ever will. Creating teams with shared goals and incentives is often essential for building collaboration during cultural wars. Becoming the "common enemy" of enemies that must work together may also be an effective tactic to break through the resistance.

SSP4: Having worked for California-based companies for over 15 years, I've frequently referred to the California Consumer Privacy Act (CCPA), which later evolved into the California Privacy Rights Act (CPRA), as the privacy standard. You might wonder – isn't GDPR considered the global gold standard in privacy? Opinions vary; some agree, while others don't. It's important to note that complexity can be the "Achilles" heel' of any security and privacy program. For organizations not engaged in business within the UK, EU, or Switzerland, GDPR may not always be applicable. The CPRA, on the other hand, is crafted with the unique legal context of the United States in mind and aligns more closely with other US laws and regulatory frameworks. Unlike the often rigid requirements of the GDPR, the CPRA offers a more balanced approach suitable for various types of businesses. Fortunately, both standards prioritize what truly matters – consumer rights. While the GDPR has set a high global standard, the CPRA provides a regulatory framework that is finely tuned to the specific legal, cultural, and business environment of the United States, potentially making it a better fit for American companies.

Target grouping

During a company merger, the privacy program in the United States had more resources and experience than the privacy program in Europe. The GDPR was going into effect in several months, and the company had many clients who would be subject to the regulation.

The project turned out to be one of the more successful projects during the merger and only had one minor misfire, which was pulling in a consulting team to examine control readiness. As expected, the report found that we were completely unprepared for GDPR, but it was issued long after our project team had begun the compliance journey.

To get things started quickly, I traveled to Europe with my privacy focal, where we could collaborate with the teams. Within Europe, many teams had suggestions on what needed to be done, but no one wanted to be accountable.

After speaking with several teams, I was contacted by an employee who reported to the Chief Financial Office (CFO) – and came to be instrumental in moving the project forward. With most of the management team in Europe, including the Chief Executive Officer (CEO) and CFO, having this new team member and her connections allowed us to:

- Bring the regulatory issue to executive management's attention.
- Ask for budget, resources, and executive support.
- Pull together a project team to work on compliance.
- Gain agreement that my team owned GDPR compliance.

Once these were implemented, we had about eight months to achieve compliance. Fortunately, my US Head of Data Privacy was instrumental in driving the project forward. Over the next eight months, we collaborated with multiple teams across the enterprise, held GDPR awareness meetings to gain employee support, and hired a privacy expert in Europe to strengthen the program.

Six months after becoming compliant, we knew the project was successful when personal data was misplaced, and the new process which included notifying the Information Commissioner's Office (ICO) within the required GDPR timeframes, was followed.

⊕ Sharpshooter Tips

- Learn about the organization and find out who the influencers are. I would not have been successful without the help of the individual directly under the CFO.
- Be accountable. If you can be physically present, do it! Being present adds credibility to the team.
- Hire and rely on competent people. My Head of US Privacy was vital in driving the project forward.
- Provide privacy training for the entire company when everyone must adhere to new or increased regulatory requirements.

↝ Misfire

- Hiring a consulting team to identify an issue is usually helpful in achieving executive support. However, if your timeline is short, you may have to try other avenues or run the program in parallel. We would never have made our timelines if we had waited for that report.

⊇ Everyone follows policy, right?

As a global CISO, I managed the security aspects of the merger with another European company. Due to the time pressure to merge the two firms, I spent a good portion of time at the European headquarters, working through implementation plans to merge the two security teams. The European team was small and governance-driven. In contrast, the team in North America was large and more operationally driven. For example, the North American team ran and managed the vulnerability scanning software and submitted findings for the IT teams to remediate. In contrast, the IT team in Europe ran their own vulnerability scanning and remediated based on the findings.

In talking with the Head of Security in Europe, he reiterated that the IT team performed vulnerability scanning and remediated findings according to policy. I asked him what standards were used to harden the company workstations, and he replied that he had spent considerable time hardening their IT systems against the Center for Information Security (CIS) benchmarks. I asked him to run a compliance report from the vulnerability scanner to understand how closely the company followed the CIS benchmarks. As he pulled the report, he exclaimed, "This can't be right; our workstations are only 35% compliant with the hardening standards we agreed upon."

Sharpshooter perspectives

SSP1: It looks like the Head of Security in Europe realized that his process wasn't working. It was clever of you to ask for a compliance scan so he could see the program's effectiveness for himself rather than you saying it wasn't working. I've never seen a compliance program work where security only has oversight, and IT uses all the tools. What works is for security and IT to work together to review the benchmark controls, evaluate the gap analysis together, and agree on implementation plans. Both teams should be able to use the tools and see the results. It's best to implement the low-risk controls first, then medium, and finally, the high-risks. Controls with significant operational concerns should be thoroughly tested before implementation. I've seen cases where controls are implemented but must be backed out due to operational impact. Thankfully, the changes didn't cause significant issues since thorough backout plans were created as part of the change management process. Due to environmental constraints, some controls are not feasible, and compensating controls must be implemented as an alternative. If there is any residual risk, it should go through the risk management process, be placed into the risk register, be accepted, and periodically reviewed for future action.

SSP2: Trust, but verify comes to mind. One of the critical movers here is to have joint dashboards and reporting on a monthly reporting cycle. We use monthly operational reviews, where the teams responsible for the security functions meet to review the metrics and prioritize tickets, enhancements, and projects. The various teams meet regularly and review the same metrics and dashboards that executive leadership reviews, which skyrockets compliance levels and remediation efforts.

Target grouping

When we merged Information Security programs between the two companies, we found that the teams in charge of patching were also managing the vulnerability scanning system but weren't following the agreed-upon standards.

Based on the compliance scans, I moved the responsibility of vulnerability scanning to the security team. Although the decision was easy to justify based on the scanning results, it was challenging to implement because it was a shift in accountability and process for the IT teams. I discussed the benefits of the change with the IT team, which included allowing them to focus on their core functions and not having to worry about the scans. Even with the scanning process shifted to my security team, it still required a lot of collaboration since the IT team continued their reluctance to give up complete control.

⊕ **Sharpshooter Tip**

- In the event of compromise through an unpatched vulnerability, the CISO is often held accountable, regardless of their control over the patching process. While a CISO may be perceived as a "big brother," it's critical to have an independent team verifying security. One of the most vital considerations is using a vulnerability scanning solution separate from the patching solution. This segregation of duties reduces errors and provides a reliable verification of the patching process, which may have ignored critical vulnerabilities.

⌐⊸ **Misfire**

- Avoid coming down hard on the patching team without understanding why patches may be missing. In this story, we found several benchmark checks had been turned off, and the signs pointed to the IT administrators being lazy. Sometimes, configurations and patches do not get pushed due to resource constraints or lack of knowledge on the part of the IT admin. By working to understand why things are not being patched, the CISO can highlight the risk and potentially get additional resources or training for the IT admins.

≥ **The insider threat is real!**

After months of working with the newly formed security and intelligence team, the board approved funding for several critical initiatives we wanted to implement the following year. One of our main objectives was to improve physical security and reduce the risk of our intellectual property (IP) being leaked by partners operating within our organization's premises. The critical project was devised to protect the company's highest-value IP, as promised to the board.

Due to the size and scale of our global company, DLP was no small effort! Our current DLP wasn't performing as expected, so we planned to replace it with a competing product.

An external consultant investigating the current DLP operation and personnel wanted to understand why the system wasn't effective before recommending a new DLP technology. The consultant discovered that a single analyst was responsible for the global DLP solution, and logs weren't being saved due to a lack of budget, which turned out to be a nominal expense compared to the overall budget.

Sharpshooter perspectives

SSP1: A DLP solution without appropriate logging for what is being protected is like having a Tesla in the Antarctic. You aren't going to get anywhere. One analyst may be suitable depending on the system design and how well it was tuned, but it couldn't have been tuned well without appropriate logging. Many CISOs make the mistake of trying to replace a tool with something new and shiny when, in reality, the existing tools would meet or exceed the requirements if appropriately tuned. I believe most security tools use less than 50% of their full functionality. I would also question whether the solution should focus on an electronic DLP system or something else since the scenario involves people who are physically on-site. Sure, you can record badge swipes to track partner and employee locations and have cameras that monitor IP storage locations. Still, I'd start with the personnel processes and physical security procedures for securely handling and storing the IP. Unless processes and procedures are in place to limit the access to IP and how easily it is accessible, no DLP system will solve the problem; rather, it will be a "we confirmed that it left" after the fact. My strategy is that they cannot walk away with IP if they don't have access to it in the first place.

SSP2: Most companies don't deploy DLP because it isn't a solution you can easily purchase and turn on. To be effective, you have to implement a significant program that includes skilled staff who can tune it correctly. Without adequate tuning, DLP can pose a liability to the company. You might find yourself in the awkward position of explaining to opposing counsel why you flagged an issue but failed to follow up on it, especially in the context of eDiscovery, which has a knack for spotlighting such oversights.

SSP3: I've been in this situation. One of the more humorous stories was watching the IT team complain that their antivirus wasn't working when systems were being compromised. After bringing in an external consultant, they discovered that IT had configured so many exceptions that the antivirus was no longer effective! In this case, it sounds like the company bought a product but didn't train or allocate appropriate resources and funding to control the risk. I would ask if DLP was implemented to satisfy a compliance checklist, which would explain the lack of resources. If they genuinely care about data loss, executive management should be receptive to allocating more resources to the project.

Target grouping

The CISO funded the initiative following a brief return on investment (ROI) discussion on testing the current DLP product with added storage before investing in new software. The DLP solution functioned as expected once the team properly configured the log storage. Shortly after, the lead engineer discovered issues that resulted in the arrest of an insider. This incident validated the team's suspicions about a probable threat and affirmed the functionality of the DLP product.

In this situation, the data engineer knew the software's limitations when the logs weren't adequate. Why would operations continue without bringing the issues forward until a consultant stepped in? The status quo was that software was installed but not fully deployed or configured correctly. A plan was in place to form a new team and spend big dollars on the latest and greatest "shiny" tool. Would they have done any better with the increased spending and new solutions the following year?

Large companies often struggle with a lack of orchestration across business silos and within complex processes. As a result, any multi-team effort requires oversight from the CISO to ensure there aren't gaps in people, processes, or technology undermining desired outcomes.

⊕ Sharpshooter Tips

- Great CISOs foster an environment where truth and transparency are celebrated, even when your "baby is ugly." This openness allows the security operations team to progress and meet their objectives. To challenge and change culture in a positive direction, a CISO must cultivate personal relationships. If in India, do "business over tea," and if in the US, take a colleague out to TopGolf or out for a drink after work. To create a strong team and safe environment, get to know and build trust with your teams by listening and connecting.
- Work with executive management to support a program such as "See Something, Say Something," similar to the Department of Defense's (DoD's) program. This program promotes a culture where individuals, including those further down in the management change, are encouraged to bring issues that affect company security to light.

⌐ Misfires

- Every sizable global company, and even mid-size organizations, suffers from "silos of excellence," politics, and operational divisions that hinder progress. If you don't address these silos, the fragmentation will result in dangerous gaps in technology, operations, and people.
- CISOs must foster an environment where every employee, regardless of position, feels safe and comfortable speaking up about operational concerns, challenges, and pitfalls.

ट Slayer of bad services

My company was undergoing a merger, so I met with the other company's outgoing CISO. He explained that due to resource constraints, they had engaged an external firm to perform due diligence on external vendors. After our discussion, I developed a roadmap of our service areas and a plan for evaluating each to identify best practices.

My current employees assumed I would select the processes from our company rather than the company we were merging with. When meeting with the management team, I stressed that, as a merged team, we needed to pick the best service for the new company, regardless of who used what service before. I interviewed the team and asked about the vendor that performed due diligence. The feedback was positive, and they mentioned that the vendor had helped them convert to an ISO 27001-based questionnaire. I looked at an example of a completed form and said, "This is impressive; how many critical vendors are we currently monitoring?"

Suddenly, the room became very silent. The lead security person said, "Well, we have contracted for them to perform due diligence on 30 vendors yearly, but they have only completed two over the past nine months. The vendor makes the excuse that they only maintain the questionnaire portal and that we are responsible for following up with the vendors and ensuring they complete the form."

Sharpshooter perspectives

SSP1: Anytime you merge operations, you must consider the economies of scale and whether you can achieve the same results with less staff. In this case, the cost savings isn't a reduction in staff but a reduction in the cost of outsourced services with the same staff. The vendor's due diligence sounded promising, but if they only supplied the questionnaire portal, that wouldn't help the resource problem. I'd review the specifics of the contract. Suppose they are contractually required to provide the portal only. In that case, I'd look for another vendor to perform comprehensive vendor evaluations and apply my headcount towards incorporating all services the managed security service provider (MSSP) owned and moving them in-house. Another reason for this is that there are many moving pieces to consolidate infrastructure and applications during a merger, and it would be much easier to outsource vendor evaluations than MSSP services during the associated chaos.

SSP2: It sounds like a lot of "security theater" going on between the MSSP, the teams, and the capabilities needed. This situation is one of those times where very transparent and high-candor communications must occur. Merging teams can mean benefiting from each other's best practices or adopting the worst. Perform a comprehensive evaluation and establish one service catalog to rule the island and only allow the best of each to remain. Kick the rest off the island. Plus, systems and services must be maintained and fully implemented. Two out of thirty over nine months? That seems weak and a dereliction of duty. Security isn't a checkbox activity, yet it seems the vendor operated that way.

Target grouping

With the merger, each security team was incentivized to show that their program was operating effectively. An inspection revealed that the due diligence vendor was underperforming.

As a CISO, you should monitor the performance of critical vendors and have frank conversations with underperforming vendors. I reviewed all the contracts and observed how my teams interacted with each vendor to ensure positive collaboration and value.

I decided to speak candidly with the vendor performing due diligence and let them know we no longer needed their services due to poor performance. They were shocked until I showed them how few vendor questionnaires they were completing compared to what was in the contract. They reiterated that it was our job to ensure each vendor responded, but then I pointed to language in the contract about them helping us complete 30 questionnaires. In the end, I let them go.

⊕ Sharpshooter Tip

- Monitor vendor performance closely. Review their contract and set up regular meetings to track and review performance. If performance lags, discuss improvement plans. If the vendor still doesn't improve, release them.

⌐⌐⌐ Misfire

- Automating vendor metrics and reports would have detected poor performance much earlier and allowed the team to address it with the vendor. If possible, put systems in place to automatically deliver metrics for review every week.

⊒ Managing the department of "Yes"

Security teams often gain a reputation of being the department of "No," which is commonly bestowed but rarely deserved. I worked as a Director of Information Security and managed the teams that performed vendor due diligence and security architectural reviews. One afternoon, I was called into the office of the CISO, and the conversation went something like this,

CISO: "The Chief Technology Officer (CTO) says that your team says 'No' all the time and that the information technology projects aren't completed on time because your team is unreasonable."

Me: "I don't recall the team saying 'No' to any information technology projects."

CISO: "The CTO says he polled his team, and your team has a reputation for saying 'No' instead of working to support their projects. Your team needs to work on saying 'Yes' more."

Sharpshooter perspectives

SSP1: This is the default perception of the information security team even before the first interaction occurs. It reminds me of my first day as a CISO role at a new company. One of the best IT Field Solution Engineers told me that security was holding up all her projects. I wondered, "How could that be when there wasn't a security program?" Security was not holding up any projects, but her perception was that my being there and implementing a security program would inhibit her ability to execute rapidly. The solution to this "No" perception was to find ways to say "Yes" and be part of the solution. In this case, we found a way to have a "project site in a box" pre-approved by security. To set up a new site and have connectivity back to headquarters, IT shipped the crates of pre-configured servers, PCs, network equipment, and satellite connections to the site. Electricity was added, and it was ready to go. The result reduced connectivity time for a new greenfield site from ten-plus days to less than three. Wins like this help make security supporters out of security foes, which changes the narrative to "Yes, Security can!"

SSP2: This is often an artifact of considering security as something in addition to a solution instead of an essential and integral part of every solution. In modern technology implementations, there isn't a solution that doesn't require security. To suggest otherwise is a complete misunderstanding of the current state of affairs. I would ask the opposite when presented with that narrative: "Why does {IT/Marketing/? } insist on creating organizational risk by ignoring the essential components of cybersecurity, privacy, and compliance needed to survive and thrive as an enterprise?"

Target grouping

The CTO had approached the CISO and stated that the information security team was halting information technology projects by saying "No" to too many requests. The information security team had a reputation for being unreasonable.

I took the CISO's comments seriously and spent time with my team to set up a tracking system that documented every risk-based question and analysis we were asked to work on. We already had processes for vendor due diligence and technology risk analysis. Still, my team was not as disciplined in documenting if someone asked us a minor question such as "Can I whitelist this email domain?" or "Can we delay patching to complete an upgrade?" The system acted as a risk register, and we discovered that we documented around seven analyses each week. After 2 years, we determined that the team said "No" 0.52% of the time. Having the facts made it easy to refute that the security team was refusing every request.

One positive outcome was that my team and I improved at quantifying and expressing risk in a way that made sense to other organizations. The team also became better at stating why controls needed to be implemented, with most risk analyses becoming "Yes, but you need to put these things in place to manage these risks." We also improved our documentation when someone in the business signed off on a particular risk.

⊕ Sharpshooter Tips

- Cultivate a culture that uses a "Yes, and ..." approach. Instead of shooting down a solution by saying "No," train the security team members to think about how they could implement the solution securely. This mindset shift led to a culture where the team members said, "Yes, and we need to put in these controls."
- Even the best information security teams will encounter reputation problems. As a risk management professional, all CISOs will eventually have to inform someone that the product or strategy carries substantial risk for little gain. I've heard IT professionals express sentiments such as "This is the most lock-downed environment I have ever had to work in; the security controls are out of control in this environment" or "I used to work at such and such a company, and they never were this strict even though they were in a more regulated industry than we are." In addition to keeping good documentation, having a good network of other CISOs helps. I'll often contact CISOs at companies where people worked, claiming they had more lenient security controls, but I discovered that most of those assertions were false. Documenting risk analyses can be time-consuming. Resist the temptation to give verbal assessments and always follow up with a written analysis. Documenting and tracking risk analyses will save you and your team's reputation and should protect you when things go wrong. Facts at your fingertips will help you cut through rumors and maintain your reputation.

⌐⟶ Misfire

- It would have been easy to get defensive and say, "Everyone hates security." This stance would have led to greater animosity between teams and soured the reputation of the information security team.

ORGANIZATIONAL STRUCTURE

Organizationally, many CISOs report to the VP or Head of IT. This positions the CISO's budget within IT and shows that the company views information security within the IT realm. In other organizations, the CISO may report to Legal, Compliance, or a Risk management team, showing that the organization views it as a risk management function.

The CISO's reporting structure and level are also essential. Many CISOs provide reports to a company board. Depending on the CISO's reporting level, board reports may be filtered through a Chief Information Officer (CIO), a CEO, or other layers. Although regulators may require that the board have access to Cybersecurity experience, gaining access to the CISO may be difficult because of the layers of management between the board and the CISO. The CISO has a challenging job of managing communications and conveying risk to senior leadership throughout the company.

2 Two weeks into a new CISO role

Taking a CISO position at a firm in the fall, I began working with the team to understand the gaps in the current environment. Organizationally, I reported to a department that managed risk and compliance within the organization. The company had substantial gaps, but the management was on board with me proposing a roadmap to close the gaps. I was acutely aware of the typical budget planning cycle: in August or September, teams begin budget planning, and in October, senior management begins to review budget proposals. Since it was already fall, hopefully my team had already been working on the budget!

I was met with blank stares in my manager's meeting as I inquired about our budget. I realized that the security team had been moved out of IT six months before I joined. I asked the obvious question, "Where is our budget?" Fortunately, one of my managers responded that our budget remained with IT. No one on my team knew what was in the budget, though. At the least, I knew who I needed to speak with to understand my budget – or lack thereof. Unfortunately, I would have to hurry if I wanted to propose new funding to close the gaps.

Sharpshooter perspectives

SSP1: One of my favorite books is "Extreme Ownership," which discusses how a team – and every person on that team – takes on the responsibility for success in a mission. It's clear from this example that the leaders in the room have some severe leadership gaps, having not championed, cared, or even known what the budget needs were or attempting to enable or "take care of business" for what is critical for a team to survive. One might excuse this leadership gap with the recent restructuring, but make no mistake: leaders lead, which is a massive failure in leadership. What are they doing, where is their focus, and what are they doing to ensure the team succeeds and moves forward?

I'd address the budget needs immediately, as that is a critical need with an urgent timeline. There is a political and alignment leadership challenge given how the teams were bifurcated and the budget left in IT. I'd be on guard to feel the lay of the land and culture that led to the restructuring, what they hope to accomplish, and the personalities involved. As I navigate that new space, I'd ensure my team understands extreme ownership, with each person measured for their leadership roles and responsibility – especially around their attitude supporting the security team's internal customers. This focus is to ensure no man is left standing alone and the team is successful no matter what projects they're working on. This mentality involves team-building exercises and a lot of 1:1 time to build trust and a solid relationship. I led the team by example, with some long hours during the first six months, to set the stage for success for this fractured team.

SSP2: The reality is, "No budget, no program." If possible, meet with the CFO and find out where the security salary, expense, and capital budgets are maintained. The security budget may be in IT's bucket but may not be broken out into its general ledger (GL) or by line item, so separating the two might be very hard. I'd take this as an opportunity to create the budget based on your years of experience, which would be my ready ask when meeting with the CFO. The best case scenario is that the CFO accepts your budget request, puts it in its own GL, and subtracts that amount from the IT budget.

Target grouping

After moving into a new CISO role, I found out that my team's budget was part of the IT budget. I met with the CIO to understand the budget. He was a little surprised that I wanted to know about the budget. During the conversation, I found out that there were several security projects that he had championed for Information Security. One of the most apparent issues was that no one in my team had been accountable for budgeting in the past, hence the CIO's surprise that someone was now interested in the budget. In addition, the CIO informed me that the budget would be submitted in 2 weeks, so I had a short time to add new items.

I then pulled the security team together to plan a roadmap for the upcoming year, including the projects the CIO had championed. I asked the managers to pull together the gaps that they had seen within the environment and worked with them to pull the projects into a roadmap with allocated resources. I worked on estimates with the managers for each project and then reviewed the results with the CIO. Although the budget stayed within IT, I could add additional items with justifications to close specific gaps. Ultimately, the CFO approved the budget so the team could move forward to remediate the gaps.

⊕ Sharpshooter Tip

- When entering a new organization, it is critical to understand how budgets operate and how they are allocated. Understanding the budget should be a priority for a CISO entering the organization.

〰️ Misfire

- When I entered the organization, I should have asked more questions about the gaps and how the program worked. The CIO knew about gaps that my team did not know about, and if I had learned these earlier, this would have helped me rapidly define a roadmap with my managers.

2 The VP of IT has already spent the budget!

The company I worked for regularly bid on request for proposals (RFP) for different services. For each RFP, a security analyst reviewed the proposal and created an estimate of the security services cost. Recently, the company won a large contract to support children's healthcare.

One of the IT architects working on the implementation pulled me aside and asked about the hard drive encryption since we had to buy specific software to encrypt the hard drives.

His first question was, "Do we need to encrypt the hard drives? After all, the drives will be on desktops that are on site."

I responded, "Yes, we need to encrypt since it was specified in the RFP and required by HIPAA regulations."

He asked, "What if we don't write any data on the local hard drive?"

I responded with a little more emotion, saying, "Well, in the RFP, they specified that we needed to encrypt the drives. So I understand; why are we so hesitant to encrypt the hard drives?"

He reluctantly replied, "Oh, the project was over budget, and the Vice President of IT has already spent the money that was supposed to be for encryption, so he wanted me to ask you what else we could do."

Sharpshooter perspectives

SSP1: First, I don't believe that data won't be written on the local hard drive. Even if the user doesn't do it, and most do, then there are processes and temp files running in the background that do. I'd quickly investigate whether using dumb terminals or virtual desktops could be viable since there would be no local hard drives to encrypt. If that doesn't work, the VP of IT has to decide whether to be personally liable for not being HIPAA compliant or overrunning the project budget.

SSP2: No encryption, no implementation. Security and compliance are NOT afterthoughts or "nice to have things." It has to be part of every solution, or that solution does not go into production. Ignoring an essential security control and breaking the law can put you into an orange jumpsuit. The VP of IT will now need to find the budget elsewhere because without meeting the proper security, privacy, and compliance requirements, there can be no ATO (authority to operate). Hardline. The lack of encryption on a local system was the root cause of one of the most significant breaches I ever managed.

Target grouping

The IT department managed project budgets, including security funding. When a program exceeded its budget, the VP of IT diverted funds from security controls to cover the deficit and, in this case, disregarded the client's contract terms. When I showed the contract requirements to the IT Architect and the VP, it became evident that the contractual obligations held more weight in their eyes than the HIPAA requirements, as the contract was a significant revenue source for the company. After discussing this, we agreed that the hard drives would be encrypted to protect the data. The architect then proceeded to encrypt the program using Bitlocker, a recently released solution by Microsoft.

Sharpshooter Tips

- Maintain good relationships with people in other units. Since I had a good relationship with the IT architect, he told me about the issue and genuinely wanted to work together to find the best solution.
- Understand that some individuals bend compliance to deliver a solution more often than they should, placing a CISO in a risky position. By showing that the contract would be at risk, I demonstrated that there were clear consequences for not following the contract.
- Understand your accountability in the situation. As a CISO, I laid out a framework in which regulatory compliance must be a foundation for security controls. Most executives agree with that framework because of the risk and fines associated with non-compliance. One approach outlines the risk of not putting in the control versus the business loss. In addition, if you find yourself in a situation where maintaining regulatory compliance is being questioned, you may want to start looking for another company.

Misfire

- My misfire in this situation was not having a better pulse on the budget. The money was spent in this case, and I had to work hard to recover. If possible, work with finance to segment the security budget from the other IT budgets.

If we get breached, we'll just fire the CISO!

I was recently hired as the CISO of an engineering firm that held several government contracts. Shortly after I joined, government representatives contacted management to point out several missing security controls in one of the engineering programs. Being relatively new, my boss convened a meeting with the directors to discuss remediation.

During the meeting, it became clear that engineers working on the program felt that the proposed security controls would prevent them from working as quickly as before. They felt that adding the missing controls would cause the program to exceed budget and prevent them from delivering on time. They recommended that our team have a conversation with the government representatives and outline the risks to the program.

The directors discussed different approaches to the issue. During the conversation, one of the directors mentioned that the missing controls put the program at risk of a security breach. Without missing a beat, my boss said, "If we get breached, we will just fire the CISO."

Sharpshooter perspectives

SSP1: Welcome to the team. Sadly, you have no job security or authority and are officially the scapegoat in a culture that doesn't value security! It angers me that CISOs are held personally responsible and liable by organizations, the Federal Trade Commission (FTC), and others when the company culture and other leaders often set them up for failure, only to use them as a shield to protect their reputation.

Migrating from DevOps to DevSecOps is a tough challenge. Developers care about developing, making things work, adding functionality, and meeting user requests. Security folks care about security, while clients care most about new features and functions. The tension is real where developers frequently feel that security requirements are burdensome, time-consuming, and undesired while ignoring the responsibility of accepting the security risk they may unintentionally create. You can win this battle by using a matrix to qualify and quantify the risk, developing clear policies and procedures, and creating a steering committee of security and developer staff championing a balanced approach.

The best methodology for a life-cycle of DevSecOps is to have the developers kick off a program or agile/scrum cycle involving the security expert(s), who then use a theory-based approach, as well as architectural, to identify security gaps, threats, and techniques, tactics, and procedures (TTPs) used by adversaries to target specific applications. Identifying security gaps enables developers and security experts to identify threats, vulnerabilities, and requirements. The steering committee then formally negotiates risk-based priorities, according to policy and procedure, and risk acceptance as they navigate what goes into the product or not. To improve culture, the lead engineer and developers on the committee must champion business requirements and a culture of accepting and valuing collaboration.

Developing a risk-based spreadsheet is another way to help move the needle on accepted risk and cultural challenges with leadership and developers. It details the "cost of an incident" specific to your company and models scenarios like including vulnerable software in a release. This approach provides a clear view of costs and risks for informed decision-making. With a financial calculator, you can map this back to ranges of low, medium, and high risks associated with threats or risks and then develop policy threshold decisions or options. For example, accepting the risk of vulnerable software for a few months is very different from accepting the risk of multiple vulnerabilities that have significant exposures and are exploited by adversaries. Let data and numbers drive the conversation and decision-making, changing the narrative from what people want or value to focusing on business risk.

SSP2: When I faced this kind of negative attitude, my mindset was, "Well, if you fire me, you better enjoy it because you will only get to do it once." It was empowering since that was the worst that could happen until recent legal changes. I thought, "If I get fired, there will always be other companies to work for." To prevent this, I started informing the boss and documenting the risks. While most managers won't sign up to accept the risk, a paper trail shows they were told, clearly knew, and accepted the security risk. In reality, the engineering team can probably modernize their processes and become more efficient while incorporating security, but will they even try?

Target grouping

Government representatives identified numerous gaps in the company's programs. When concerns arose about potential breaches due to vulnerabilities, my boss casually mentioned they would just fire the CISO (me) if an incident occurred. Hearing this sentiment repeatedly, I devised a two-part strategy. Firstly, I would raise awareness among engineers about the client's requirements. Secondly, I actively pursued other CISO positions, sensing that the company viewed me as a potential scapegoat and lacked genuine concern for the client's interests.

I collaborated with my team and engineering to meet with government representatives to discuss project status. In most meetings, the reps were firm but reasonable to work with. The engineers were frustrated as they didn't understand why they needed to be involved but kept stressing the new controls would slow down their processes. After a painful journey of pulling the engineering team along with us, we eventually implemented the required controls.

I stuck it out with the company, hoping things would improve, but unfortunately, the same situation occurred on two other projects. I began to realize the issue was a lack of management support. I was hopeful things were changing when management accepted a roadmap I developed to enhance controls, aligning them with our client's requirements. However, I learned that upper management didn't fully support the roadmap. The head of engineering didn't embrace the plan either, so my team and I struggled being stuck between polarized groups. After going through the same painful journey a couple more times, I found another CISO position and moved on to another company.

Epilogue: For the next several years, my former team members would periodically contact me. One told me, "You know all those projects you were working to implement? After several breaches, management finally realized they had security issues and started to remediate the vulnerabilities. A lot of the projects you had sponsored finally got implemented."

⊕ Sharpshooter Tips

- Before working at this company, I discovered the previous CISO had left after only six months. I later heard that most CISOs rotated out of this company every six to twelve months. Repetitive short CISO tenure is an indicator of poor management support. Before joining a company, I recommend discovering why the previous CISO left.
- Don't hesitate to move on from a company where upper management consistently puts you in the position of having to argue with clients about their security requirements. Often, this resulted from a manager signing off on the security requirements without understanding the cost or implications. If possible, request the security team review all contracts before signing.
- Another CISO once told me he was always "one click away" from losing his position. He was referring to the fact that someone could click on a phishing email, triggering a security incident and him being blamed, which sadly happened, and he was fired. A completely negligent CISO can and should be fired if a breach occurs, but if you are reading this, that's probably not you. I've known several CISOs that have negotiated a clause in their employment contract that states the firm must provide six months' notice before termination to provide a buffer if they are blamed for something and fired. Also, you should find out if you are covered by the company's "Directors and Officers (D&O)" or similar liability insurance. This insurance protects you if you are blamed for an incident.

⌐ Misfire

- During the interview, work to understand the culture and the support from upper management. If I had known more about the culture, I probably wouldn't have left my previous company for an organization that didn't take client security controls seriously.

2 Why it's important to C.R.A.S.H. as a CISO

It was a windy and cold day with definite signs that a storm was coming. I sat outside near the fire pit at a favorite coffee spot with a colleague who happened to be in town. We talked about what it was like being the top security executive in our companies. At one point, he took a long breath and said, "I'm not sure how much longer I can do this. I feel like my life is a combination of Groundhog Day and that guy who continually pushes the rock up the hill only to have it roll back down (Sisyphus). I'm mentally burnt out, and the hits keep coming." Thinking about that momentarily and being the ever-supportive friend, I said, "Welcome to the club. It never gets easier, but it doesn't have to be as hard as you're making it." That earned me a profanity, but I quickly followed up with a bit more clarity: "What makes a difference is not the challenges, but your attitude towards them. It would help if you adopted a mental model that supports your sanity and long-term success in the role. You need to be a 'C.R.A.S.H.' focused CISO."

Sharpshooter perspectives

SSP1: I can relate to your friend who is "mentally burned out" since my objective as a CISO was to never have a significant compromise. To me, that was the only way to win. I had a hard time accepting the statement, "It's not 'if' you're going to be compromised, it's 'when'." That may have been acceptable in some companies, but in my previous organizations, a compromise would have catastrophic implications for the safety and well-being of the United States. I agree that if a nation-state has a target, they will eventually compromise the organization if it's connected to the internet. Protecting the most valuable assets with segmentation and having a business continuity plan is essential. I can't wait to learn more about C.R.A.S.H. because I still lean towards the premise that a significant compromise isn't acceptable, making the CISO position very stressful.

SSP2: Self-care is critical as a leader, especially in high-stress situations and a demanding work environment. I learned early on to be intentional about my self-care. Little things add up to making a big difference. For example, most meetings we have on video conferencing are scheduled for an hour. I set mine to be 50 minutes by default. Why? I don't want back-to-back meetings; I want a few minutes of breathing room between meetings. If I don't manage this proactively, I'll run ragged from meeting to meeting, struggling to get a break between calls, emails, pings, and everything else.

Self-care also comes into play with understanding my needs for success and setting aside scheduled focus time each week for strategic work to get ahead of the crazy reactive work of everyday life. This focus time helps me be more efficient and balances reactive and proactive activities for tactical, strategic, and operational outcomes. It improves my sense of excellence and helps me move the ball at work!

In my personal life, I have to remind myself, as a driven, passionate, ambitious person, to maintain a solid work-life balance where I prioritize sleep, nutrition, exercise, spirituality, and other matters so I bring the best "me" to work.

By prioritizing self-care and being emotionally aware, I can bring my best self to work every day and ensure my needs are met. What small actions can you take each day to reduce stress, prioritize self-care, and grant yourself the freedom to step away, walk outside, rest your mind, and get a head start rather than getting swept up in the rat race and risking burnout?

SSP3: In one of my CISO jobs, the CIO came to me and said, "The CEO is asking the successful leaders of the company to put together a one-page essay about what is making your area successful, and you have been recognized as a successful security leader."

I wrote a paper titled "Staying positive when the odds are stacked against you." I think it surprised the CIO that the CISO role can be remarkably stressful, and needless to say, it never got presented. One thing I feel is important is understanding where that stress originates from. If the stress is worrying about getting compromised by an external attack, that is part of the job and the puzzle you have to work through to defend the company. If the stress comes from upper management not being on board with your program, you should figure out if you're at the right company.

Target grouping

Anyone can be a CISO. It is just the title of a job. Plenty of people will give you lists of skills, frameworks, experience, and credentials you need to be a CISO. Those are all fine and good, but there is one thing above all else that will determine whether or not the job will make or break you. That is your mindset. Suppose you want to be a successful executive leader in information security. In that case, having the right attitude and mental models in place is vastly more important than having any particular experience, training, or tools. Lasting success as a CISO comes from a generative and resilient mindset, which will give you everything you need to navigate the situations you encounter. I like to call it a "C.R.A.S.H." mindset, which stands for Curious, Resilient, Aware, Strategic, and Humble.

The C.R.A.S.H. mindset

Curious: Stay curious and always look for ways to learn from the situation regardless of the circumstances. Approach each situation with a mindset focused on learning and be ready to question your current understanding. This openness encourages growth and helps you gain the insights needed to find your desired solutions. Curiosity will drive you to investigate deeper and look behind the obvious. It will help you explore beneath the surface and enable you to find the usefulness of everything. Curiosity also drives focus and eliminates fear. Many times, you will be gifted with opportunities that can be terrifying. Being curious about them will drive more proactive and positive action to discover why and how, while fear drives reactivity and adverse action, which is not what you want during a cyber incident.

Resilient: Things will never be perfect and never get easier. Get over it. The key is to become stronger each day so that dealing with challenging situations and obstacles becomes more manageable. It doesn't matter how many times you get knocked down; it only matters how many times you get up. Each time you get up, you get stronger. As you become stronger and more resilient, knocking you down is more challenging. Take the hits. Learn the lessons. Be better for it. Strive to get a little better each time, and the compound effect will surprise you.

Aware: Know your world. Be aware of the current environment and the threats you must deal with. Know your most important assets and give them the appropriate amount of time and effort. Measure the impact of your actions and adjust accordingly. Rinse and repeat. This resilience includes all aspects of your program.

Strategic: Know your organization's goals and your role in enabling them. Architect your program to fulfill that mission. You are not "just" a security resource. You are a vital capability and organizational executive with expertise and accountability in information security. Play your part in the success of the larger mission. This strategic mindset means you will need to learn the business side of the house and the goals of the other groups within it. Be the partner you wish you had.

Humble: Approach every situation with an open mind and a willingness to learn. Admit when you are wrong or don't know something. Seek opportunities to improve yourself and your knowledge. Surround yourself with people with diverse perspectives and experiences, and truly listen to what they say. Stay grounded, and don't let ego or arrogance cloud your judgment or prevent you from seeking or accepting help. You are not infallible, and even experts make mistakes. Mistakes are learning opportunities if you maintain humility. Stay hungry to improve while also being confident in your abilities. Have patience with others, and focus on growth over blame. A humble attitude enables continuous improvement and stronger connections.

⊕ Sharpshooter Tips

- You can't control the circumstances you are given, only how you respond to them.
  - Success and failure are less about the circumstances we are given and more about how we interpret and use them to our advantage, regardless of the details.
- Secure your ground first so you have the stability to support the people and organizations that rely on you.

⌐᷒ Misfires

- Not adopting the C.R.A.S.H. mindset leaves you vulnerable to situations that will negatively impact you personally and professionally. Here are a few:
  - Burnout – Not caring for yourself first depletes your energy and resources, leading to physical, mental, and emotional exhaustion. This neglect prevents you from being able to serve others effectively long term.

- Enabling harm – Being unaware of your limitations can cause you to make mistakes that negatively impact or even harm those you are trying to serve.
- Loss of credibility – Overextending yourself could cause lapses in judgment that undermine your reliability and trustworthiness as a leader or mentor.
- Increased stress – Disregarding your physical, emotional, spiritual, and mental needs places additional stress on your body and mind, hindering your ability to make calm, rational decisions.

Chapter 2

Leading the way

Practical leadership and governance for CISOs

In the complex world of cybersecurity, the Chief Information Security Officer (CISO) can be paralleled to a skilled marksman maneuvering through challenging terrain. This multifaceted position requires a comprehensive understanding of technology, the ability to implement controls without disrupting business operations, and the diplomacy to recommend mitigation measures while balancing security and operational risks, even if they may temporarily limit business activities. This chapter is dedicated to the pivotal aspect of leadership and governance, providing comprehensive guidance and practical advice to empower CISOs to excel in their roles.

YOUR NUMBER ONE PRIORITY: LEADING PEOPLE AND THE ORGANIZATION

The CISO has many priorities, with the top four in order of importance: people, processes, technology, and controls. However, their primary focus is leading their team and shaping the direction of the organization they serve. This focus can be likened to the role of a seasoned marksman on the competition field. Just as a sharpshooter rallies their team, a CISO must inspire and guide people within the organization. Their leadership involves:

- Understanding the individual strengths and weaknesses of team members.
- Fostering a culture of cooperation.
- Navigating the dynamic landscape with finesse.

Like a marksman at the forefront, guiding and motivating fellow competitors, a CISO must also be an unwavering leader, ensuring the organization's overall mission and security objectives are always on target.

⊇ We can do ten!

My first computer security position was a special assignment from an Executive VP at a Department of Energy (DOE) facility. The DOE was unhappy with

DOI: 10.1201/9781032720500-3

our company operating this large nuclear materials generation site and had fined us $1,000,000 for computer security failures. One of the security violations was failing to wipe personal information from hard drives before being sold and shipped to China. The hard drives contained the Social Security Numbers (SSNs) of DOE personnel, so Congress got involved to provide special oversight, and several Congresspeople visited the site to highlight the seriousness of the issue. My "temporary" job was to address the one hundred computer security deficiencies discovered during an audit within 6 months to prevent our company from receiving another fine, which was expected to be much larger and potentially cause us to lose the security portion of the contract. Like other DOE contractors, our company was in both a hiring and a budget freeze, so my initial request for resources was denied. Fortunately, twelve people were in the Computer Security organization but were already working overtime to keep up with issues and demands. How could I approach one hundred deficiencies meaningfully in 6 months?

Sharpshooter perspectives (SSP)

SSP1: One at a time. There is no easy way to get around the fact that demand can only be satisfied with available capacity. If there isn't capacity, the demand level doesn't matter. It won't get done. In these situations, the best scenario involves stack ranking the vulnerabilities and affiliating them so similar work can be done together. Then, start work and produce an ongoing progress report related to the requested timeline. When the Executive VP sees the completion date beyond the desired date based on resources applied, the reprioritization of projects and wants (read resource demands) can occur to bring that event horizon into line with the desired end state. 100% is 100%, and an individual can't give more than 100%. The job of leadership is deciding what tradeoffs need to happen to get the most important work done within that 100%.

SSP2: This is a remarkably tough issue. Without resources or a budget, the only mechanism I know that may work is prioritizing vulnerabilities by risk, which the executive team should approve. Then, assign the remediation to teams outside the security team and have the remediation tracked at the executive level. The status would need to be reported weekly at an executive level. The challenge is to ensure that each unit's executives are accountable for remediation and have an executive sponsor at the highest level who will ask why initiatives are falling behind schedule. As the schedule slips, this will likely cause the executives of each area to be tasked with remediation to request resources and highlight the lack of resources issue.

Target grouping

Success was achieved by prioritizing the list, being bold in getting top brass buy-in, finding creative solutions, and working tirelessly to deliver.

My first step was to take the "100 List," completely understand each item, and estimate the following for each one individually.

- Perceived importance to DOE.
- Labor required to develop and implement a solution.
- Materials and contract labor cost to implement.
- Time to implement.

I prioritized the list and identified ten items I thought would be a very high priority to DOE and could be implemented with existing staff and equipment. I met with the Computer Security team that I had been assigned and showed them the prioritized list. As we discussed the top ten, someone said, "We can do ten!" which became our rallying cry. There was only one big catch. The team agreed, but would DOE? I scheduled a meeting with the DOE Site Manager, who is responsible for a site of 30,000 people, to discuss the proposal. It was a bold move, but the best option, since the issue was on his staff agenda each week. Of course, there was a lot of brass in the meeting, but the DOE Site Manager made it a discussion between us, where he asked, "Can you do these ten in the next six months?" My response was, "Yes, we can?" He responded, "If you complete those ten in the next six months and have a plan to resolve the others, I will consider this matter closed." We were thrilled to get his support.

To delete the personal information on all the hard drives, we removed them from the excess systems, using the chain of custody handling processes, and shredded them. The team was highly motivated because resolving these ten items would mean they were no longer in the DOE spotlight, and they would repeatedly remind each other that "We can do ten!" We resolved all ten issues and met with the DOE Site Manager 6 months later to close out the findings. When he walked into the meeting room, he saw small containers of shredded hard drives sitting in front of him. He picked one up and said, "Well, no one can get information from that!" Turning to his DOE Security Team, he asked, "Are we doing this with our systems, and if not, why not?" There was a meek "no" response, and his attention quickly turned to them. He stated the matter with our company was closed and asked all contractors to leave. My leadership in addressing this matter led to the Computer Security team recognizing me as their leader. The team requested senior management assign me as their Deputy Director permanently. This successful project transformed my "temporary" assignment into an exciting career.

⊕ Sharpshooter Tip

- Taking on and conquering monumental tasks is a great way to accelerate your career because it shows you are a leader, adds high value to your company, and demonstrates that you are the go-to person when there is a challenge. Approach with caution, but do so with enthusiasm and positivity.

- Consider a whole array of alternatives and present those with the highest probability of acceptance by the decision-makers. Be open to their decision, be flexible, and adjust as needed for maximum impact.

⌦ Misfire

- A common misfire with monumental tasks is approaching it with the attitude that "there is no way." There is always a way to success.

PERSEVERANCE, FINDING A WAY, AND ADAPTING

To excel as a CISO, you must "learn the business." This knowledge is more than understanding security standards; it's about getting inside the heads of your peers across the business – especially those in nontechnical positions. Understand what drives them, how they communicate, and their priorities. The technology language may be foreign to them, but your ability to translate security concerns into business terms will set you apart. Build relationships and engage in regular conversations with key stakeholders to bridge the gap. Your role requires building relationships across departments. Effective relationship-building will break down barriers and facilitate collaboration and understanding. Ascertaining the motivations and concerns of multiple departments, such as marketing, finance, and operations, will enable you to align security objectives with your company's broader business goals.

⌕ You want a new classified program in a week?

Having been in a computer security leadership role for several years as a DOE contractor at one of the DOE's largest weapons sites, I was familiar with the classified and nonclassified teams and programs. Operations had become routine except for adopting controls for modern technologies like the internet and Wi-Fi. Then, the unexpected happened. On Wednesday, a classified security incident occurred within the DOE complex (not associated with my DOE site), shutting down all of the DOE's classified computer operations. No processing could occur until DOE Headquarters (HQ) approved the new program. However, DOE didn't provide a schedule or guidance on how to correct the perceived deficiencies. On Thursday, the head of the DOE's Tritium program, responsible for producing and delivering Tritium (a rare and radioactive isotope of hydrogen used in nuclear weapons) to the Department of Defense (DOD), called and was concerned that a shipment was scheduled for the following Tuesday and that shipment had to be shipped. The site had never missed a shipment since its inception 50 years ago. However, a shipment can't be made without recording and processing the classified data within the system, and it couldn't be processed due to the shutdown. I had to figure out how to allow the shipment out the door in 5 days. I had no authorization to process classified data, so I had to get a program approved by HQ and have a system certified for processing in 5 days, including the weekend. A classified computer security program had never been approved in less than a month. Senior management was already starting to communicate there would be a missed shipment and blamed DOE HQ for it. Do I accept this situation as "That's how the cookie crumbles," or do I adapt and find a way to make it happen?

Sharpshooter perspectives

SSP1: I assume that the government DOE stakeholders in the program have already met with the contractor about this. If not, that would be a first step. Depending on the DOE flexibility, they could allow some of the resources in a related program to be used under certain constraints. I am interested in how this one turns out because, most of the time, everything stops in a situation like this.

SSP2: Although I am inclined to find a way to "adapt and persevere," I must go with the cookie crumbling in this context. Cyber threats are dynamic, and systems and processes must be proactively developed around dealing with them in a compliance-heavy environment. That means planning for them before-hand and often because of missed objectives. A "seat of the pants" approach is never a good idea when dealing with classified systems. If it's down, then it's down. Sorry, not sorry about that. Ad hoc adaptation is extremely risky and can lead to even more significant harm than a missed SLA.

Target grouping

Well, there are 5 days, and I'm not about to miss a Tritium shipment in the "Cold War" era if I can help it. I contacted the head of Tritium and told him I would have a program written on Saturday so it could be reviewed on Sunday and sent to HQ by Monday morning. He agreed that if the program were solid and would work, he'd send a recommendation for support and approval through the operational side of DOE to HQ. I contacted the Classification Office to make sure that what was going to be proposed could be written in an unclassified version to distribute. I knew that the other DOE sites were in the same mess of being unable to process classified information, so I called each one and told them I would have a program written by Saturday if they could review and support it. I solicited and gained their input during each of those calls. They agreed to be available, and if the program would work for them, they would send letters to the DOE supporting the program by Monday morning. Of course, I coordinated with my team and management to ensure they could also review the program.

After skipping sleep for a couple of nights, the detailed program was writ-ten to address the deficiencies noted in the incident, and the program was distributed on Saturday morning. It received broad support and was approved without reservation by everyone. Letters were transmitted to DOE HQ by all DOE sites by Monday morning. HQ approved the program by the close of business that same day! A system was certified to the new program on Monday night, and Tritium shipped as scheduled on Tuesday. Within a year, the "Cold War" was announced as being over, and I'd like to think that in some small way, making sure this shipment of weapons materials helped achieve that national goal.

In retrospect, getting a compliance program developed, approved, and implemented in 5 days was monumental. The team's dedication continued a 50-year history of consistently shipping on time.

Here are some pointers from the trenches on how to adapt, find a way, and persevere.

1. First, you must believe that the cause is worth your extraordinary effort, that it can be done, and that you are committed to making it happen.
2. Don't listen to those who say it can't be done because they probably don't want you to do it for selfish reasons. If they are in a position of influence over what you will be doing, try to get their buy-in. Otherwise, put them in the loser category, ignore them, and keep moving forward.
3. Align with others that can benefit from your actions. They can be your support structure when things appear bleak and a great resource when you need an idea or assistance.

⊕ Sharpshooter Tip

- Take on the difficult challenges you believe are the most important and require you to adapt, find a way, and persevere. These significant successes will be the most rewarding in your career.

⌐═ Misfire

- Choosing the wrong course of action because it appears too hard or the outcome is uncertain will always be a misfire since it will result in nothing of significance being achieved.

⊇ You Can't change our identity process

While working in Computer and Information Security (C&IS), I worked at one of the Department of Energy's (DOE) largest locations. Multiple offices were involved in getting someone approved to work at the site, each with independent systems. The main three offices involved were Personnel Security, Operations Human Resources, and Construction Human Resources. I was engaged because C&IS required someone to be uniquely identified before computer accounts and systems were assigned. Depending upon whether it was a hire for operations, construction, or a temporary contractor, there could be multiple paths. Still, all potential workers were generally processed through Personnel Security first and then by the associated HR office. With

25,000 full-time workers and thousands of contractors, keeping track of everyone and ensuring they were uniquely identified was challenging. Using SSNs to identify everyone was standard then, but there had been recent guidance to discontinue their use. The objective was to shorten the time required to assign computer accounts and equipment, which could take 2 to 3 weeks due to the sensitive nature of the processing. Still, that process could only start once HR entered an employee into their system since that was the first indication to C&IS of a new employee. There were other processes, such as nuclear exposure tracking across the site, where employees needed to be uniquely identified. To complicate things even more, DOE recently stated that SSNs for their employees could no longer be used after 60 days. DOE employees had to be part of the Personnel Security System, where every person had to be vetted and assigned badges. A meeting was held among Personnel Security, both HR departments, and C&IS to determine a solution. Both HR departments were adamant that they should be the ones to control employee identities and felt the identities of Operations and Construction employees should be unique to each organization. C&IS said their work needs to start before the HR identification assignment, so the solution would not solve the delay in assigning computer accounts and systems. Personnel Security said it wasn't their issue since government regulations required them to use SSNs to complete their work, and they transferred approved requests to the HR organizations. What could be a workable solution here? How could SSNs be removed from all systems while uniquely identifying everyone?

Sharpshooter perspectives

SSP1: One entity – one identity. Robust identity-proofing mechanisms are essential; uniquely identifiable information is core to meeting that mission. SSN is a common identifier but doesn't have to be the only one. Having a mechanism for an alternate identifier is critical to navigating the issue described. For instance, if offshoring is allowed and non-US citizens don't have an SSN, how would you accommodate that population? I am still trying to work through challenging problems today in a holistic and rational manner. I am interested in seeing where this one goes.

SSP2: I worked for a company that used a unique but sequential set of ID numbers assigned to employees; for example, your ID would be something like ID24709. Assigning and resolving any argument on a person's ID was easy. The issue was that someone could test a password on the remote access site against most of the IDs in the company by using a script to iterate all the potential IDs. The company also had passwords expire every 90 days. During a penetration test, the testers wrote a script to iterate over all the IDs and tried a couple of passwords, such as "Spring2023!" and "Summer2023!." The testers succeeded in compromising 1% of the company's accounts. The testing demonstrated the issues in easily guessed IDs and how asking for a 90-day password expiration pushed individuals to pick predictable passwords. Although the problem of assigning unique IDs seems on the surface like an easy problem to solve, it is a complex issue that requires good coordination and creative solutions so as not to make the identity and access management system more insecure.

Target grouping

Someone external to the situation suggested a solution that was the catalyst for everyone making concessions, and a successful unique ID process that lasted over 20 years was implemented.

The group was fortunate that a person in the meeting was visiting from another Personnel Security Office. She saw that each group was focused on their needs and that a deadlock was occurring. She said, "I saw this problem once before, and someone proposed that Personnel Security assign a unique number to everyone evaluated, and that number should be used in every system. They couldn't agree to do it, but I think this group can, and it could solve your identity problems." The way she phrased the solutions was fantastic, especially subliminally challenging the group to implement a solution another group couldn't. Technical Security quickly said they could assign a unique number of five characters long so that up to 99,999 people could be assigned a unique identity. Both HR departments asked that a variable be added to the front to identify the type of employee: "O" for Operations, "C" for Construction, and "T" for Temporary. They would keep the same number, but the variable could change for the same person if they changed work status, which was uncommon. C&IS said they could immediately begin the account process once the unique identifier was assigned so that accounts and computer systems would have a greater chance of being ready when the employee arrived. C&IS coordinated with IT, and within 60 days, unique IDs had been assigned to everyone, and every computer system had been changed to accept the new ID. This situation demonstrated that sometimes it takes someone from the outside to provide a creative, nonbiased solution by "thinking out of the box." Also, if someone is wondering, the ID was not allowed to be part of access credentials.

⊕ Sharpshooter Tip

- Security personnel, including CISOs, tend to be very factual and structured in reaching and applying solutions. Consider involving others who may propose more creative ideas when developing solutions.

⌒⌐ Misfire

- Each group focusing only on their immediate problems and not considering the bigger picture and long-term solutions can be a misfire, resulting in missed opportunities for improvement.

COMPANY-WIDE RISK MANAGEMENT PROGRAM

The CISO is in the risk business, and having a formal, company-wide risk management program is essential to any cybersecurity program. Just as marksmen meticulously prepare their shooting equipment, a CISO must craft a risk management plan to safeguard the organization. A CISO's mission is to ensure that all departments align their focus and actions with the bullseye of risk mitigation. Much like a seasoned marksman ensures that each competitor follows a consistent and safe approach, a CISO's program unifies the entire organization, emphasizing the importance of collectively hitting the security and risk management target.

Your risk management program should be closely aligned with the broader business goals and the overall risk appetite of senior management. When cybersecurity aligns with the company's mission, it becomes a driving force for its overall success. Security standards can be used to gauge risk to the organization. Choose the standard that aligns with your company, and if there is no apparent applicable standard, choose one, such as the Cybersecurity Framework from the National Institute of Standards and Technology (NIST), that aligns with any business and get started. Any standard is better than no standard. Anywhere there is a gap between your program and the standard, it is a potential risk.

≅ We already know our risks

I was asked to help a company establish a security program as a Virtual CISO (vCISO) on a part-time basis. This company was not only in shambles from a cybersecurity posture, but IT was also in rough shape. I agreed to help, so I first researched their business to determine what regulations and standards would be applicable and the basis for establishing the security program. It was clear to me that they were a "health care provider," one of the three types of entities covered by the Health Insurance Portability and Accountability Act (HIPAA). One of the first security requirements within HIPAA is a risk assessment. I dusted off my HIPAA controls spreadsheet containing all the controls and designed to identify gaps between the company being evaluated and the regulation. I decided to meet with the General Counsel (GC) and discuss what they believe are applicable regulations and control standards. After all, it is their company, and they need to understand and speak to their security program. I was shocked by the discussion. The GC was unsure that they were a "health care provider" and seemed to have every counter argument as to why they were not. One of the most unexpected things was his response to conducting a risk assessment when he said, "We already know our risks." The theme was that HIPAA is too burdensome, compliance will take too many resources, and be too expensive. The argument included things like our legal structure, which makes HIPAA not applicable, and our provision of limited health services. This attorney was in complete denial when any layperson would look at the business and conclude, "You are a health care provider." So, how do I approach this to persuade the GC that HIPAA was applicable and that we needed to conduct a risk assessment to determine compliance gaps to be used as the basis for a security program, including corrective action plans to close the gaps?

Sharpshooter perspectives

SSP1: This is a challenging issue. One way that I have worked through this is to pull up information on competitors, obtain agreement that they are competitors, and then show that the competitor has HIPAA duties. Another way is if the company is a service provider and I can get copies of contracts, particularly a business associate agreement that the company has signed, I will use it to demonstrate a business requirement. This story has hooked me, and I am interested in how it comes out.

SSP2: When the organization's GC is adamant about it not being under HIPAA when, from a reasonable person's perspective, it would appear to be so, I would phone a friend. And by a friend, I mean outside counsel, the organization's insurer, or a regulator to get confirmation. As a CISO, you must be willing to go against the grain sometimes and get the confirmations and assurances you need to be certain. People are fallible, CISOs and lawyers included, and you need to be ready to stand your ground, ask the hard questions, and accept the ultimate decision on risk tolerance made by the executive leadership team. Document any concerns if those decisions fall on the wrong side of logic and rationality.

Target grouping

Ultimately, the GC accepted that the company needed to be HIPAA compliant. The risk assessment used to show HIPAA noncompliance and other business risks became the company's risk register, which the GC used quarterly to manage the company's risk.

I knew I would not convince the GC in that meeting, so I asked for the opportunity to review the business and another meeting next week to review the basis for a security program. This was to buy time while I put together a presentation to include options for establishing a security program aligning with company operations, approximate costs, approximate timelines, and business risks for not having a program. I also coordinated with the new CIO to get their buy-in on the correct approach. We jointly gave the presentation a week later, and the GC, while not entirely buying into HIPAA compliance, agreed to fund the completion of a security and privacy control HIPAA risk assessment.

The spreadsheet used included a row for each security and privacy control. It also included columns for the current state, identified gaps, corrective action needed, timelines, and costs. Since no program existed, significant gaps would require long-term programs to be developed and implemented. We identified tactical, strategic, and operational risks to address short- and long-term gaps.

- Tactical risk management is concerned with day-to-day activities and short-term objectives. It deals with immediate and specific risks that can impact the organization daily.
- Strategic risk management focuses on the organization's long-term goals and direction. It involves assessing risks that could affect the achievement of strategic objectives and the organization's mission.
- Operational risk management bridges the gap between tactical and strategic levels and is concerned with the processes and activities essential for the organization's day-to-day functioning. Operational risks can have both immediate and long-term consequences.

The first 4-hour session resulted in the risk assessment being about 80% complete, with several actions requiring follow-up with the GC, VP of Operations, and others. In addition to the HIPAA controls, the team identified several risks within the environment that needed to be resolved. This initiative started a cascading effect, with IT personnel identifying additional risks and placing them under a "Potential Risk" tab as they were identified. I used this initial risk assessment to develop a presentation of the risks faced by the company and a prioritized list for addressing those of the highest concern. This got the GC's attention and convinced him that the company needed to be HIPAA compliant. The company-specific risks also received a lot of attention, and he asked to review the entire risk assessment. This resulted in the risk assessment being the basis for the company risk program. The risk assessment was converted to a risk register that was used to track progress on corrective actions, log potential new risks for evaluation, and record risks that had been resolved. The IT team met monthly to review and update the risk register and met with the GC quarterly to report progress and to adjust focus as needed.

⊕ Sharpshooter Tip

- Pick a standard for your company and use it to perform a risk assessment. The key is starting and then continuing the process, which is the win; it's not always how accurate the initial risk assessment may be. Continuous reviews and ongoing changes will make the resulting risk register more accurate for the organization.

⌁ Misfire

- Failing to anticipate the GC's pushback on the need for HIPAA compliance was a misfire. I should have recognized this was a possibility and have been more prepared since they hadn't done anything towards HIPAA compliance in the past.

⌷ Direct network external connection

I had been the highest-ranking security professional at a large engineering and construction company for about 3 months when my role was expanded temporarily to include responsibility for end-user computing and the service desk. About 1 week into the additional responsibilities, I learned of an "emergency" ticket for my Executive VP. His company-supplied home laptop wasn't working. I opened the ticket and read, "Laptop no longer works. When the laptop is turned on, it stops at the login prompt. It didn't do this before. When turned on, it never asks for a login and would connect directly to the company network. This needs to be fixed immediately because this is the laptop the Executive VP's wife uses." You've got to be kidding me! Someone at the service desk must be playing an April's Fool joke on me since it is April 1. A short while later, I got a call from the Executive VP asking why his home laptop no longer works. He is serious and wants it fixed ASAP. I explained that we recently had implemented a new security control that requires passwords to expire. This control was needed to protect the company from account compromise and potentially losing high-value data. He was not pleased and wanted the password on the laptop changed. He also wanted to review the new security program since he thought it negatively impacted the business. Where did I go wrong? Due diligence was done, including a risk assessment of the environment and creating a plan to address the technical gaps required for a sound computer security program. I was working with Operations to implement these technical controls. How could the Executive VP not understand that the steps taken were to secure an insecure environment?

Sharpshooter perspectives

SSP1: This is a world of wrong supported by entitlement and personal convenience. It's likely a company policy violation when his wife, who's not an employee, uses and accesses corporate resources. Gather some allies in HR, Legal, and higher on the food chain, then address the issue head-on. Allowing a culture of convenience over the right thing is dangerous, and depending on the industry, the company could end up in legal action or face criminal charges.

    SSP2: This is likely to be a painful conversation. Do you have any allies who know the VP you can talk to before the conversation? The most likely way to change this situation is if an unfortunate event, like a security incident or a new regulation, makes it mandatory. On the other hand, if the VP's wife needs a laptop, one solution might be to buy her a computer and allow her to use it off the company network.

Target grouping

In this scenario, I had focused on the massive technical control deficit and failed to implement a cybersecurity awareness program. The awareness program should have started with a high-level briefing for senior management on the risks, gaps, and corrective actions being implemented while also requesting their support. Once senior management approved, we should have educated IT personnel on the details associated with the risks, gaps, and corrective actions. Then, we should have deployed the general awareness campaign for end users, explaining the new controls, why they were required, and any change expected in the end-user experience.

Here are some tips to assist with implementing an effective awareness program.

1. Keep senior management and those people of significant influence briefed on your security program plans and inform them when new controls or significant changes will be implemented. They are much more understanding and can even push back when they receive complaints if they understand what is occurring and why.

2. Expand the security knowledge of IT personnel specific to their position so that they consider security as they perform their day-to-day activities. They should be aware of controls implemented through the change control process, but that isn't enough. Visit their all-hands meetings and explain the changes in depth. Provide knowledge-based articles that will assist them in appropriately responding to user issues and requests associated with implementing new controls.

3. Combining vendor-supplied security awareness programs with targeted emails about company-specific threats increases end users' awareness. The frequency and depth of the training usually depend on the organization's readiness and openness, which can evolve.

⊕ **Sharpshooter Tip**

- Cybersecurity awareness at all organizational levels is required and critical due to the sophistication of spear phishing. For example, security software may not detect a phishing email coming from a compromised account, but the individual user can.

↝ **Misfire**

- When you come from a technical background, it's natural to prioritize technical controls, especially when a security program is a mess. Technical controls are crucial and need to be implemented. However, for true success, the controls must be explained through a robust security awareness program that spans the entire business and is endorsed from the top.

POLICIES AND PROCEDURES

Every cybersecurity program must have policies and procedures detailing how security controls will be implemented. These documents should align with the implemented standard, serve as a reference point for all employees, and ensure everyone knows their role in maintaining security. Policies can be grouped into two broad categories: technical policies that govern IT and administrative policies that apply to the entire organization. Technical policies are typically only distributed within the IT department to avoid confusing general users.

⊇ Yes, you need policies and procedures

When I begin new cybersecurity leadership roles or provide consulting, I usually find the organization's security program immature. Their programs lack a robust cybersecurity framework, so one must be built from the ground up. The scenario has been consistent across all industries. One indication of a nonworking framework is the lack of policies, procedures, and standards. The absence of established policies, procedures, and standards poses a significant challenge. The documents can be written to meet applicable compliance requirements, but in many cases, the organization will be in noncompliance immediately upon document approval. The documents are needed to ensure that consistent practices are followed and are the basis for starting the maturing process. Legal counsel has told me that if a breach occurs, it is better not to have policies, procedures, and standards than to have and not be following them. So, this appears to be a catch-22. The policies, procedures, and standards are needed to mature the program and be written to meet compliance requirements. However, in most cases, the organization does not meet and must implement programs and processes to reach compliance. So how can policies, procedures, and standards be written to compliance and implemented while avoiding the misfire of immediately putting the organization in noncompliance with its documents upon approval?

Sharpshooter perspectives

SSP1: This is a worthy problem to discuss! I have had the same issue in working on a merger. I put one person in charge of merging the policies of both companies and allowed him to outsource policy creation. What came back was almost a word-for-word set of the ISO 27002 standards. I had to be very patient and work with him on pulling together policies that reflect what we did as a company. I inquired about certain areas of the policy and if we were doing them, and he said, "We should aspire to be doing these." We discussed what we had in place versus what we planned to implement, leading us to carefully use words like "shall," "will," "must," "should," and "guideline."

SSP2: Happens all the time. You need a policy but are afraid to write the correct one because you won't comply with the new one if it is ratified. The solution is to write the proper policy and build an adoption period into the process that establishes a rational and achievable glide path to compliance. For example, if the specific policy is about establishing strong identity assurance in the organization, then make sure you split the policy portion from the current requirements and procedures adopted by the organization to be compliant with said policy. The policy says you will ensure strong identity assurance; procedures reference current practices that are sunset when the "new" more robust control practices are adopted.

SSP3: I'm a huge proponent of "structured for success" with a balance for all outcomes, including legal compliance and application wisdom. Some may use the legal concerns of accountability not to create policies and procedures, forcing organizations to be ad-hoc and people dependent on operations, leading to inconsistencies and immaturity. The best balance starts with policies and procedures that holistically address critical controls highlighted in ISO/NIST and similar frameworks without being overly specific. This approach ensures excellence at 5,000 and 25,000-foot levels on what matters most while leaving the more challenging and detailed components out to avoid excessive specificity, legal liability, and operational burdens. Consult legal counsel to ensure this meets your legal appetite for your organization!

Target grouping

Establishing policies and procedures was the first step in creating a foundation for security that protected the company's digital assets. With the right approach, collaboration, and reliance on industry best practices, we successfully built a resilient cybersecurity program that adapted to increasing threats.

Writing a full complement of policies, procedures, and standards that meet compliance while actively implementing programs and processes that comply with the documents can be achieved. Here is what worked and what didn't.

What Worked

1. To ensure your program complies while these documents are implemented, focus on policies, procedures, and standards. Understand the difference between each when developing a comprehensive set of documents. Be aware that at least one person will get wrapped around the axle about what documents are called and, even more so, how they are formatted. They are a distraction unless you can get them to focus on the overall objective, not the terminology. Define how you will design the documents and focus on consistency. Here are generally accepted definitions of the three types:
   a. Policies: High-level documents outlining the organization's cybersecurity goals, objectives, and principles. They set strategic direction and define what needs to be achieved.
   b. Procedures: Detailed, step-by-step instructions for specific security tasks or processes that provide a practical guide on effectively implementing the policies.

c. Standards: Technical specifications that set the requirements for various security aspects, such as password complexity, encryption protocols, and network configurations.

The interplay between these elements is crucial. Policies provide the overarching goals, procedures define the processes for achieving those goals, and standards detail the technical requirements. In some cases, like for small organizations, combining all three in the same document may be beneficial.

2. Include a statement regarding the status of your program in the Information Security Policy (ISP). It is acceptable to note that the program is under construction, and the organization will not be compliant until fully implemented. Most people recognize that security is a journey, and compliance can't be achieved overnight.

3. Perform a gap analysis between your practices and what the policies and standards require. Develop realistic corrective action plans to close the gaps along with timelines and implement them. This is where procedures will be identified and implemented since they support implementing the requirements within the policies and standards.

4. State in the header of the policies and standards the date of approval and an expected date of compliance based on the correction action plans. Adjust the expected compliance date if something adverse happens, impacting the scheduled implementation date.

5. Avoid adding items not required in the policies and standards. For instance, don't say that you are wiping drives to NIST 800-88 requirements when your process is physical destruction by shredding and recycling.

What Didn't Work

1. Ignoring the issue of immediate noncompliance. Auditors thrive on finding discrepancies in your practices and documentation.

2. Including overwhelming detail in the policies, procedures, and standards. While detail is important, too much can lead to complexity and confusion. Policies need to strike a balance between providing clear guidance and being practical for employees to follow.

3. Skimping on training. Having policies in place was a great start, but employees were still at risk of making mistakes without proper training and awareness programs. We needed to complement the policies with a comprehensive training program.

⊕ Sharpshooter Tip

- Security policies are necessary to meet generally accepted security practices. They are also required to defend a company that has experienced a breach.

☞ Misfires

- Including requirements in approved policies that will not be implemented.
- Focusing only on writing the policies rather than verifying the policies align with the organization's business objectives, tooling, and enforcement controls.

౽ People will quit!

Often, security teams become the bearer of unwanted news when a regulation or client requirement changes how people work. One of our customers established a new security framework and communicated that we needed to comply with their new standards. They sent an audit team to evaluate us, which resulted in several deficiencies. Being our biggest customer, no one at the company wanted to see them unhappy. One finding was that there was no policy on "session timeout after a period of inactivity." Having recently joined the company, I talked about the policy with one of my senior architects. He informed me that the engineers refused to allow that policy to exist. I was incredulous, but I worked with him to write a policy stating that a session would be timed out if there was an hour of complete inactivity. I then approached engineering management, saying that our largest client had asked for the policy and that we wanted their feedback on what had been written. I was shocked when he said," If you implement this policy, people will quit!"

Sharpshooter perspectives

SSP1: A great business case was made regarding why a timeout policy was needed. However, history at the company caused a response of "people will quit." I would ask probing questions, discover concerns related to that history, and then shape the proposal to address those concerns. If there isn't an exception policy, then I would create one and let those who express concerns have the ability to submit for an exemption using an Exception Request form. The exception process creates a way to show their concerns are heard and a way to say "Yes" when there is a valid argument for not having a timeout. Note that most people will be satisfied that their concern was heard, and only a few will have a concern valid enough that they will make an effort to complete an exception request. If an exception is approved, the policy should state that corrective action will be taken within a year or resubmitted for re-approval.

SSP2: Unfortunately, I get this one more than I would like. System timeouts are in place for the safety, security, privacy, and compliance of the organization, the workforce, and customers. They are not there for the sake of annoyance. That said, workflows and environments need to be considered. A 60-second timeout may not be reasonable, but a 5-minute one could be. Anything beyond 15 minutes, though, is more of a workflow process issue than a technical one.

Target grouping

The session timeout policy resulted in one individual quitting the organization. I left the organization approximately 2 weeks after the policy was implemented. Here is what led to my quitting that company.

I met with the engineers and listened to their concerns so I could put an exception policy in place for legitimate reasons. They talked about test cases they performed over Secure Shell (SSH), and the tests could run for several days. The Security Architect on my team pointed out that even over SSH, they would have to have zero activity for over an hour for a session to timeout. The engineers stated that they were concerned about any timeout, and I reiterated that the customer required us to have a policy on session timeouts. I asked them what they felt was reasonable. The engineering team stated that 24 hours of inactivity would be a reasonable amount of time. I agreed we would make an exception for SSH to have a timeout of 24 hours. The engineering team asked that the policy go through the Engineering Control Board (ECB). That particular ECB meeting would become infamous throughout the company, and in the process, I learned a lot about the company culture.

For the ECB meeting, I had prepared two slides. The first slide was a problem statement with the customer's requirements, and the second was the SSH exception. I never got to my second slide. Engineering management profusely stated that the policy would make people quit when they couldn't do their work. Additionally, they claimed the security team had a secret agenda to implement controls that would soon make it impossible for the engineers to perform their work. I had been briefing the executive management regarding the audit and controls and assumed that the VP of Engineering was on board with the audit results since he had never objected to them. However, in the ECB meeting, it became clear that the VP had not communicated the audit results to his directors. During the meeting, the VP remained silent. Fortunately, a different VP corrected the engineering team, stating that executive management had been briefed multiple times on this audit and that there was no secret agenda.

Although the ECB meeting was painful, there were several positive outcomes:

1. The meeting made me realize that upper management was not aligned and that there would be considerable resistance to implementing the required security controls. Throughout my tenure with this firm, I encountered areas where the controls did not meet regulatory scrutiny, ultimately pushing me to resign.
2. Executive management dissolved the ECB and established a new board with more senior managers. The executives realized that the other directors had the potential to cause client and revenue loss. In fact, customers began withholding payment due to security issues they had found in our programs.
3. The policy was put in place with the SSH exception.

I sent the policy to the engineering team for comments, and they recommended only grammatical changes but nothing substantial. Our largest customer accepted it during the follow-up audit.

⊕ Sharpshooter Tip

- Listening to the engineers and implementing the exception within the policy allowed my team to address the customer's security concerns and explain the business rationale for the exception.
- Lawyers have guidance on when they should withdraw from representing a client, including for ethical reasons. As a CISO, one should establish those ethical boundaries and be prepared to resign when the organization they represent crosses that boundary.

⌐⤳ Misfires

- I underestimated the organization's negative culture and the power of the engineering team within it.
- Going into the conversation, I had assumed that upper management had informed their directors and managers of the audit and the organization would make a good-faith attempt to comply with the requirements established by their largest customer. Never assume!

INFLUENCING WITH OR WITHOUT AUTHORITY

To effectively influence and communicate the importance of cybersecurity, the CISO should provide regular updates to executives, the board of directors, and stakeholders. Use short anecdotes and real-world examples specific to your organization to illustrate the evolving threat landscape with potential risks and impacts on the business. Develop security metrics, tie them directly to the business whenever possible, and publish them using your company's cadence for business metrics. Don't just present statistics; tell a story. Use stories to convey the successes and challenges your team has faced. Highlight how security measures have prevented or mitigated potential breaches. If your company doesn't use metrics, starting the metrics program would demonstrate your leadership and management skills.

⧈ What keeps you up at night?

As a CISO for a global company, I frequently had to speak to multiple boards across the organization about the company's security posture. The company was broken into over 24 legal entities, so I created a template that I updated monthly to provide updates to each of them. The template was simple and had information on the team, top risks and statuses, and where the company fared regarding security with its competitors. The final two slides of the presentation were used to provide each board with something that directly affected them.

Once, I provided a slide regarding a fake website that had been used to impersonate the entity's business and the takedown status. Although I always included risk items with status, the most interesting question I got from the board was, "What keeps you up at night?"

Sharpshooter perspectives

SSP1: Will I have a security team when I wake up? The demand for security professionals continues to rise, which is more significant than supply. Therefore, we continue to see salaries increase between 8% and 14% a year, depending upon the study consulted. HR subscribes to the old school thought that average salaries should rise no more than 4% a year, and if they band together with like-minded HR departments across multiple industries, they can jointly hold salaries artificially low. How do you hold on to and show exceptional employees that you have their best interest at heart when you can only offer raises less than 4% when they are facing inflation between 7% and 10%? All the while, the company's revenue increases at the same rate as inflation or more. It doesn't take long, and employees feel the pressure of no longer being able to pay monthly expenses. That forces them to look for higher-paying jobs no matter how much they enjoy the job or like you as a leader.

SSP2: As mentioned, talent acquisition and retention are the big ones. In my experience, there is still some cognitive dissonance within organizations around properly enabling and incentivizing the role of a CISO. They want it to have the responsibilities of the "true" C-Suite without the accompanying agency associated with being a part of it. When the "Chief" of an entire organizational function is in name only, it is hard to justify the salary and retention action types for the resources that report through them. Another "keep me awake" concern is discovering that basic controls I assumed were in place are not, allowing a significant security incident to occur. I'd love to trust more, but I still have the regular experience of people choosing convenience without regarding the impact on others.

Target grouping

The board asked me, "What keeps you up at night?" I responded that my biggest concern was staff retention, not technical risk since the security staffing market had a much higher turnover than other industries. They looked concerned, so I informed them that my team's retention rate was much better than the industry average – which I should have mentioned earlier. I also realized I hadn't discussed my concerns with other executives, such as the CFO, and should have done so before the meeting. Fortunately, our excellent retention rates and market-aligned salaries satisfied the board's concerns.

⊕ **Sharpshooter Tip**

- As a CISO, you must be extremely prepared for board meetings. Usually, board materials are prepared weeks in advance and diligently reviewed by everyone. If possible, get to know each board member outside the meetings and understand their cybersecurity experience and concerns so that they can be addressed before or during the meeting, reducing many questions during Q&A.

↝ **Misfire**

- Identify, analyze, and understand all your risks across people, processes, and technology before the board meeting so you'll be ready to respond thoughtfully and avoid appearing unprepared during discussions.

⊇ Can I really measure cybersecurity?

In all my security roles, senior management has always wanted to know the current level of security risk and whether it was going up or down. They wanted me to measure the risk reduction from their security spend to ensure a positive ROI. At first glance, calculating the company's risk exposure is simple. We used the risk register to identify all the known risks and assigned risk levels (number) and added those up, which provided the current risk. That can be tracked month to month, and if the curve is headed down, the risk is improving, and if the curve goes up, the risk is increasing. Risk today can also be compared to last year when evaluating the program's effectiveness over the past year. But wait a minute, is that truly an indication of the real risk exposure and the program's effectiveness? New vulnerabilities may be discovered outside the program's control but may have been present for months. One incredible CISO I worked for wanted to make these standard Key Performance Indicators (KPIs) more accurate. We managed a large complex of buildings and occasionally discovered additional equipment with vulnerabilities that had been there for months. There were also vendor-supplied systems that had vulnerabilities detected that had been present for extended periods. Once the vulnerabilities were discovered, it meant the risk KPI had been inaccurate for previous months, as these unknown vulnerabilities were not included in the KPI. Should we go back and update the risk metric to the best estimate of when those vulnerabilities were introduced into the environment to be more accurate, or should we add the risk to the current month and leave the past KPIs alone?

Sharpshooter perspectives

SSP1: Do not change the KPI risk for previous months. Instead, provide a complete story of why the risk changed. The issue is that risk will constantly shift with new vulnerabilities or findings. Changing the previous KPI does not benefit the company except to show how it's improving its metrics program.

SSP2: Keep the previous KPI ratings and adjust the current ones. The previous KPI determinations were made with the best information at the time, and all decisions were dependent on those metrics. Retroactively changing the "facts" used in the past to make decisions that cannot be remade (no time machines here) would remove the traceability of why those decisions were made. It's best to fall on your sword once, then make the significant shift and use it as a point-in-time reset for the future.

Target grouping

We decided to update the risk KPI retroactively to when the newly discovered vulnerability was introduced into the environment. This retroactive update made the metric more accurate for previous months and showed a more accurate risk profile for the complex. However, updating the metric became exponentially difficult each month as additional risks were discovered.

I find measuring the success or failure of a cybersecurity program to be one of the CISO's most difficult tasks. The ultimate measure is how much a breach or compromise impacts the business. Of course, the smaller the impact, the better, with the level of acceptable impact influenced by the business's risk appetite and willingness to invest in security controls. Here are some basics that the CISO must understand about KPIs or metrics:

1. Keep the measurement displayed simple. Ideally, a KPI should only measure one thing and contain no more than three data points on the same graph. I've seen KPIs so complicated that in every meeting where they were displayed, the discussion was about determining what was being measured rather than instantly knowing what was being measured and the performance. When I worked for a company that won the Malcolm Baldrige Award for performance excellence, only one item per KPI was allowed.
2. Make collecting the data and creating the metric as simple as possible. If the data is too hard to collect, you may find that tracking a more straightforward metric and using the time saved not collecting data may be better spent elsewhere in your program.
3. Only measure what you can and will act upon. The only reason for measuring something is if you will take corrective action based on the data.
4. Be willing to change what you measure. Stop measuring if you've had a metric for 2 years and never acted on it. Instead, look for something that needs your attention and creates a meaningful KPI.
5. Share your metrics with the people who influence the metric and make the metric feedback as close to real-time as possible. In general, people perform better when they can see the results of their actions. People will naturally focus on what is being measured.

I like to develop KPIs based on the focus of the cybersecurity program. If unpatched systems are one of my most significant risks, I would create a metric showing the percentage of systems with up-to-date patching. After the patching program matures and the metric shows the patching percentage is consistently met, I'd consider dropping that metric and focusing on another significant risk, such as vulnerabilities detected and remediated. Metrics are meant to be dynamic and to change as your program matures. Of course, you may need to have some standing metrics that don't change often to satisfy those in authority, and if so, no worries. Automate them as much as possible and ensure they are consistent and accurate.

Automated metrics are essential when performing routine tasks to reduce the drain on scarce resources. They can also enhance the accuracy and efficiency of the monitoring, data collection, and reporting processes. Here are some key factors to consider when automating metrics.

1. Use the measuring and charting tools already available in your systems, including firewalls, security information and event management (SIEM) systems, vulnerability management tools, and other security orchestration and automation platforms.
2. Develop scripts to automate data collection, visualization, and distribution of KPIs to key stakeholders.
3. Consider dashboards that provide quick data visualization and the ability to drill down into areas of interest.

⊕ Sharpshooter Tip

- Security is all about implementing controls. Metrics are a management control, so use them.

Misfire

- Making metrics so complicated and time-consuming that they become a burden rather than a beneficial management tool.

Playbooks are not for gamblers

As a consultant, I was asked to assist with implementing a SIEM system. The organization had outstanding leadership and had done an excellent job evaluating and purchasing the SIEM sized for their needs. The system was hosted on-prem and was supported by the company infrastructure team. The SIEM had already been integrated and received logs from almost all the planned systems, including network devices, servers, and the intrusion prevention system (IPS). They had determined that internal security personnel would run the incident management program exclusively unless an incident became too large. There was also a written incident response procedure, which appeared to be current, minus a few needed name changes. So, we had the SIEM, most of the required data being collected, an incident response procedure, and personnel were available. Still, none of these resources were actively being used to detect and respond to potential incidents. What was missing? As the consultant, what would I recommend and help implement so there was an active and effective incident detection and response program?

Sharpshooter perspectives

SSP1: This can be a tricky situation to analyze. You must dig down and find out the issue. Are the personnel not working together, or is the SIEM untuned and bombarding the team with data? It could be personality conflicts, training issues, playbooks taking too long, or personnel being unable to handle all the load.

SSP2: Another factor can be culture. A vendor asked me, "Why are you putting in a SIEM? If it is for compliance reasons, we have a much cheaper solution." I explained it was to manage incidents and find problems before they got out of control. That surprised him, and he said that many companies he knew put in a SIEM only to meet consolidated logging and log review compliance requirements through SIEM rules.

SSP3: Well-defined incident detection playbooks and regular security monitoring practices appear missing elements for an active and effective incident detection and response program. While the organization has invested in a SIEM system, integrated it with various data sources, and has a written incident response procedure, these components alone do not guarantee an effective program. As the consultant, I would recommend the following actions.

1. Develop and implement detailed incident detection playbooks that specify how to detect, analyze, and respond to various security incidents. These playbooks should include specific use cases, detection rules, and response procedures tailored to the organization's environment and threat landscape.
2. Establish a monitoring process to continuously review the SIEM data and alerts. Define roles and responsibilities for security personnel to review, triage, and investigate alerts regularly, ensuring that potential incidents are identified promptly.
3. Train security personnel on playbooks and the latest threats, attack techniques, and tools to enhance their incident detection and response capabilities.

Target grouping

Playbooks were written and implemented with input from all the key players, resulting in a consistent approach to each type of incident. We missed some needed information and steps, which were added as lessons learned from the used playbooks.

Playbooks describe what triggers an event and what action to take. Playbooks associated with a SIEM system are documentation and procedures that guide a security team to effectively use the SIEM tool to detect, investigate, and respond to security events. These playbooks streamline the initial event response process, ensuring every event is treated consistently and efficiently. Similarly, a football team develops playbooks with defensive plays to stop the other team's offense. A playbook should have the following information.

1. Identification Information: All playbooks should have a unique identification, including a descriptive name. The revision and date should be included so the most up-to-date playbook is used.
2. Overview: Describes the intent and context of the playbook.
3. Event Indicator: Explains what triggers the playbook, such as the unique event in the SIEM.
4. Event Priority: Explains what level of priority this type of event should receive.
5. Initial Analysis: This section describes how to perform an initial assessment of the alert, including validating its accuracy and relevance.
6. Detailed Investigation: Outlines detailed steps and data for investigating the incident, which may include:
   a. Gathering additional context and data.
   b. Identifying affected systems and assets.
   c. Analyzing the scope and impact of the event.
   d. Data specific to the incident type being investigated. Example – List Whois websites for researching IP blocks associated with external system attacks.
7. Containment: Explain how to take immediate actions to mitigate the incident, such as isolating compromised systems or deactivating compromised accounts.
8. Escalation and Reporting: Defines criteria for escalating incidents to management and what to report.

The playbook is not meant to be an incident response procedure (IRP). The IRP is a document typically initiated after the playbook results are reported. A playbook is written for a small team of analysts to determine if an incident has occurred so that initial containment action can be taken, and an IRP is written to include the whole organization and to manage an incident from start to finish. Here is a typical playbook diagram. Consider using flowcharts in your runbooks. Most will find a diagram quicker and easier to follow than if it is in text only (Figure 2.1).

The internet has excellent resources for creating a comprehensive playbook program. The Cybersecurity and Infrastructure Security Agency (CISA),

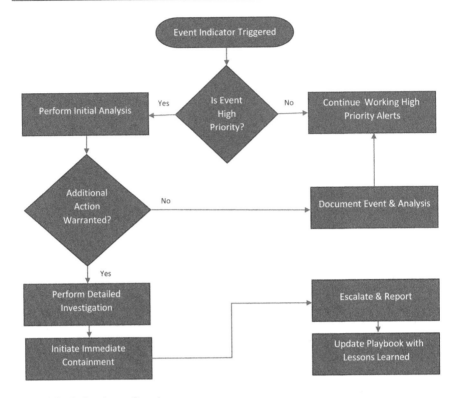

Figure 2.1 Playbook use flowchart.

located at cisa.gov, has several helpful publications. Vendors are also good resources for helping develop specific runbooks. Microsoft provides details on many incident types, including phishing and malicious applications. See learn.microsoft.com.

⊕ Sharpshooter Tip

- To get started, list all the events the SIEM is designed to detect. Prioritize the list and write the playbooks in order of priority. When developing playbooks, they don't have to be perfect. Diligently create runbooks from the best information available and continuously update them with new information and lessons learned.

⌐➔ Misfire

- While the goal of a good playbook is to have all the information needed to analyze the event, don't overburden the reader with so much information that the playbook is hard to navigate. Instead, use appendices or links for noncritical details so the user can view them at their discretion.

LEADERSHIP AND BOARD AWARENESS

Effective CISO leadership includes keeping senior management and the board informed. Keeping the board aware of cybersecurity matters is crucial for aligning cybersecurity with the overall business strategy. The CISO should regularly provide comprehensive briefings to the board, ensuring they understand the evolving threat landscape and the organization's defense mechanisms. By translating technical jargon into business terms, the CISO helps the board make informed decisions about risk management and investment in cybersecurity initiatives. This proactive communication fosters a culture of cybersecurity awareness and underscores its strategic significance, ultimately enhancing the organization's cyber resilience. When everything cybersecurity-related appears to be running smoothly, sometimes it may be challenging to get on the board's agenda. Be creative and find reasons why you need to present. One example is the ISO 27001 standard, which requires that senior management receive routine security briefings.

⮂ Your proposal is approved

Seeing the rapid changes in privacy laws and associated increasing client expectations regarding privacy, I determined that as CISO, I could no longer address security and privacy. This meant a Data Protection Officer (DPO) must be hired. Since it was midway through the yearly staffing cycle and this would be considered a senior position in the company, I needed to convince the board of directors that this position was needed and that a necessary hire would occur with expediency. I followed the required process to get on the board's agenda and was lucky to make it on the next agenda 1 month away. I started assembling data regarding how privacy was expected to significantly impact our business, including recently passed and proposed international privacy legislation, proposed state privacy legislation, client expectations, recent privacy organizational actions by our competitors, and the impact of privacy compromises on companies within the industry. After assembling this information into an eight-page presentation, I made an informational page about each board member, including their picture. I listed how each was expected to react to privacy concerns and the proposal to hire a DPO.

I knew about two-thirds of the board and thought that one-third would support the proposal without question and the other third would be very skeptical. I made time to speak to each of them and explained the situation using language that would resonate most with them, and I spent most of the time listening to their perspective. I also solicited and got their perspective on how the board members I didn't know would react. After these meetings, I felt that the proposal had the support of each person. With the newly acquired information, I met with one of the board members I didn't know and obtained their support for the proposal, though they made it contingent upon how another board member would vote. Wow, that was a win since I knew that board member supported the proposal! I still didn't know one of the board members, but I knew a few people who did and favored the DPO hire. I met with them and asked for their input, hoping they might speak with the board members. Unsolicited, one said they would talk to the board member about how this position would help meet their needs.

Based on all the feedback, I adjusted the proposal to five slides, with the first being the bottom line and ask. I included additional slides as an appendix with all the detailed information for those board members who are much more analytical. While not a requirement, I asked that the presentation be distributed to the board several days before the meeting. I was allotted 15 minutes for the presentation, so I practiced it, hoping to make it a discussion more than a one-way information push. It was the day of the presentation. Had I prepared enough? Was I providing the correct information? I was convinced that hiring a DPO was in the company's best interest, but could I convince the board accordingly?

Sharpshooter perspectives

SSP1: Data Protection is a hard sell because most companies consider it a sunk cost. Even with regulations such as GDPR, many companies roll data privacy into the information security role. You need information security controls to maintain privacy, often the driver for security controls. After reading the story, I realized the approach was sound and the best way to ask for a position.

SSP2: This reads as an excellent primer for addressing issues with the board. One of the essential items is the need for information symmetry before ever discussing it in the recorded meeting. Nothing should be a surprise when it comes up on the board agenda. If it is, your chance of receiving a negative response goes up dramatically. You are doing it right when properly preparing the participants for the meeting. Even "hard" asks are more straightforward when you follow this model.

Target grouping

Within 2 months, a DPO was hired. In subsequent board presentations unrelated to privacy, they told me how happy they were with the DPO hire and the resulting privacy program.

I was sitting outside the boardroom 15 minutes before my presentation time. I had dressed in upper business casual, electing not to go with a coat and tie since that is not the norm for the board. I was asked to go into the board room, so I took a deep breath and exhaled slowly as I entered. I surveyed the room, and everyone was smiling and seemed to be in a lighthearted mood. As I connected my laptop to the presentation system, the Chairperson stated, "I want you to know that your proposal is already approved so you can give your presentation, but it isn't necessary." Understanding that the board always has more on their agenda than they have time for, I stated, "Thank you for the approval to hire a DPO. I'd like to confirm that hiring can begin immediately." The Chairperson said, "Yes, hiring can begin immediately, and I would like to be involved in the final interview." I responded, "Of course, it would be extremely helpful for you to be involved in the interview. Since there is no need to go through the presentation and to respect your time, is there anything else security-related you would like to discuss?" There were a few questions and answers, and with half of my time remaining, I exited successfully!

Here are some practical tips for board and executive presentations:

- Remember you are a small fish in a large lake. Boards have many issues to address, and yours is probably not the largest one unless you are giving an update on a significant security incident. Therefore, make your presentation short and concise, and don't try to explain why your position exists.
- Get your point across in the first five slides. I once had an Executive VP who wanted every presentation to be five or fewer slides, so I learned to get to the bottom line in three slides.
- Get to know each board member. If there is no opportunity to know them personally, learn everything you can about them to anticipate their perspective on your presentation.
- Talk to each board member about your presentation before the meeting to get their perspective and buy-in. You can tailor your discussion to what resonates with them based on your research.
- Align your presentation to the strategic objectives of the company. Boards will respond more positively to proposals that help achieve a strategic objective.
- Tailor your presentation to the board's language. Ensure you meet all their presentation requirements and talk to others who have had success speaking to the board. Some boards are very rigid, while others are much more flexible. You must learn their standard mode of operation and adjust accordingly.
- During the presentation, if you feel they won't approve your proposal, don't take it personally; instead, look for a way to revisit it later. I've said things like: "It seems the board would like more details on our risk poster, so I'll research some additional data and, with your permission, revisit this proposal at the next board meeting." If your presentation can be an ongoing discussion rather than "one and done," that is a huge win. Don't take your proposal personally; just try again if it isn't approved.

⊕ Sharpshooter Tip

- Always, always, always make your board presentation shorter than the allotted time. Boards are notorious for running behind schedule, so if you can finish your presentation early and help them get back on schedule, they'll see you in a positive light.

🔫 Misfire

- The biggest misfire with board presentations is the shotgun approach, which sends pellets (topics) everywhere instead of using a sniper rifle that shoots a precise bullet. Keep your presentation focused and on target, and write it so a seventh-grader can understand it.

2 Systems restoration success or failure

When I was the CISO for a $5 billion company, the IT management team became concerned about their inability to restore IT services fully should a major catastrophe occur to the systems in their data center. There were backups made to tape, and the tapes were sent offsite, but no one was certain the correct systems were being backed up or that they could be restored. At the time, tapes were the predominant median for backups, and having a full infrastructure complement wasn't feasible. The IT management team presented the concern to senior management using the company's risk characterization system, requested funding to resolve the situation, and received approval. The team appointed one person as the Systems Restoration Lead, overseeing the overall program, from backing up the correct data to testing the restoration during an incident. She conducted a business impact analysis (BIA) and classified systems based on criticality, ensured that backups were being performed with the newly created backup policy, and procured the backup site services that had warm standby for the most critical systems and room for cold standby for less critical systems. She identified sources for all the systems and telecommunications that would be needed, identified IT staff responsible for the restoration, and wrote a detailed 768-page restoration plan. It took 6 months to plan the first restoration test, which included restoring all critical systems and installing all the hardware and operating systems needed for the less critical systems. The team tested the plan and found that 50% of critical systems were restored, communications from the backup site to critical offices were operational, and about 30% of the needed hardware and operations systems were fully installed. Detailed activity notes were taken, noting areas where restoration efforts could be more efficient and where work stoppages were encountered due to resource deficiencies. The Systems Restoration Lead felt that the test was a failure since not all the test objectives had been achieved, but was it? How should IT leadership characterize this exercise to senior management, who were concerned about the amount of funds allocated and spent?

Sharpshooter perspectives

SSP1: The best time to plant a tree was 20 years ago. The second best time is today. I would say this project appears to be a success. You don't know what you must work on until you see where you are. The results of the efforts give an excellent narrative about where resources need to be allocated and spent.

Any downside would likely be on the ego of those who had given assurances on the efficacy of the recovery program before the test. Take the results and plant the tree today to get the shade and fruit you need down the road.

SSP2: The project appears successful because systems are now prioritized according to business criticality, and a roadmap to improve recovery can be pulled together. This should result in a conversation with senior management to determine if additional funds should be allocated to improve the backup and recovery process. Senior management may be upset at the expense. Still, depending on the nature of the business, that would be short-sighted because the exercise indicated that, in a disaster, the company would not be able to restore all systems fully. The exercise appears successful and can be managed as part of the company's overall risk framework.

Target grouping

Senior management was impressed with the results and commented that the risk level had been significantly reduced. They considered the risk reduction investment to have been well spent.

The IT team obtained benchmark information from a technological research and consulting firm and learned that less than 25% of companies had a comprehensive system restoration plan, and only 30% of those could fully restore critical systems within a 1-to-2-week time frame when using warm standby facilities. IT leadership concluded that their first attempt was outstanding based on this information. The team recommended testing on an annual basis and using the lessons learned from the previous test to improve the restoration levels each time. During the next test, every critical system was restored except one database that had corrupted backups, and 100% of servers and operating systems were restored from cold backup systems.

⊕ Sharpshooter Tip

- An individual should be identified and given authority to develop and manage a comprehensive backup and restoration program. This role includes oversight of backups, restoration plans, and restoration implementation.

Misfire

- The biggest misfire associated with system restoration is not testing frequently to ensure the ability to fully restore.

BUDGETING

Understanding the organization's budgeting process and aligning the cybersecurity budget forecast and planning effort with that process helps the CISO obtain the funding needed to meet its security objectives. Knowing sound budgeting practices, such as detailed expense tracking, accurate estimates, and consistent compliance with finance requests, will earn you goodwill with those responsible for allocating funds. Always be prepared to justify your budget requests. Creating a story that includes gaps in your program and recent security incidents coupled with cyber threat intelligence (CTI) is a great way to provide visibility into your funding requests for senior management. It helps your initiatives become part of their strategic priorities. While no one wants to experience an incident or breach, don't let one go to waste. Immediately following a breach is the best time to ask for funding since senior management has just experienced pain and, at that time, will be more willing to approve funding to prevent additional disruption to the business.

⮂ Yes, I got budget approval!

As a CISO, I have always been faced with maximizing the benefit of my security program with limited dollars and constantly having to justify large expenditures to protect the organization while competing with revenue-generating projects. At one national company I worked for, there was a "high" risk where all data processing was done in data centers in the same city, and the secondary data center could not carry the entire load if the primary data center failed. This risk needed to be addressed, which meant a significant investment was required for a secondary data center, but there was overwhelming pressure to keep the budget flat. The annual business continuity exercise was upcoming, and I needed to develop the scenario. What could the scenario be so it would highlight the risk of the data centers being close together? Could I arrange it to highlight that a significant portion of the company was outside the headquarters city, and they would lose the ability to operate if there was a citywide catastrophe? Could I arrange it so that prominent players in the company would see the issue and propose upgrading and moving the secondary data center?

Sharpshooter perspectives

SSP1: If you can arrange to have the most senior executives in the organization, up to and including members of the board, play a part in the exercise, the organizational risk and impacts will be much easier to explain when it comes down to allocating funds. Make it interactive and express both the options and the constraints that each leader will have in the case of such an incident. Sometimes, all it takes is for the "captains of the ship" to change course on something to viscerally show how powerless they are in a critical incident if the proper preparation hasn't been performed. When risk is felt and understood at an organizational level instead of just the departmental or function level, the levers can more easily be pulled to eliminate resilience theater and create a resilient reality. A secondary data center that can't handle the total capacity of the primary data center is not a resilient reality. It is a fancy offsite backup and additional capacity station pretending to be a data center on stage for the satisfaction of an audience.

SSP2: If you can lead executive management in the right direction but have them reach the best conclusion, thinking it was their idea, you will have the most success. Involving them in the tabletop exercise is paramount. Including licensing and capacity into the scenario helps make the case because they must decide which portions of the business will operate at total capacity, which isn't possible.

Target grouping

Senior management recognized the risk as extremely high and approved the budget. They expected a fully redundant, geographically dispersed data center to be built during the next fiscal year, which was completed on time.

I had recognized the risk of having both the primary and secondary (I will be gracious here and call it a data center, but it was more of a glorified data closet) data centers located in the same city, only miles apart. It had always been that way, and the secondary data center needed to be a minimum of four hundred miles away to substantially reduce the impact of catastrophic weather events, widespread power outages, and regional telecommunications interruptions. Senior management realized the risk and wanted a data center that fully supported running operations if the primary data center and headquarters were unavailable. Plans had already been drafted for a secondary data center, and costs were quickly placed in the budget and approved. Senior management had already decided on the city location, wanting to put the secondary data center in a city where we had a substantial presence that matched our plan. The plan achieved a fully redundant data center in a different region, substantially reducing the security and business risk of losing most of the business applications needed to run the company. The business has not lost access to a data center, but if it does, it will continue operating with little to no interruption.

Here are some tips for creating scenarios where influential individuals will recognize the problem and generate solutions that feel like their own ideas.

1. Present the problem indirectly, from different perspectives, over time, and make the problem so easy to understand that a fourth grader can see the issue.
2. Highlight the areas of the problem or issue so that they think about how it might impact their portion of the business and how it will personally impact them.
3. Blend in areas of best practices and related items reported in the news.

All CISOs face a barrier to obtaining budget approval for projects and staff. While senior management may support security, they have their constraints and may not be able to fund security to the level needed. There are always budget limitations because of conflicting priorities. During my career, I've had to find creative ways to explain security program needs to senior management so that they fully understand the security risk and can appropriately evaluate it along with business risk. One way of doing this is structuring your incident response and business continuity exercises to highlight gaps in your program. Most CISOs already understand where there are gaps, so the exercises are more about educating those who can influence the budget. Structure the exercises so they can go through a logical thought process and reach the conclusion you wanted them to come to. Magically, the solution is their idea, so they own it, and what people own, they will normally support.

⊕ Sharpshooter Tip

- Structure incident response and business continuity exercises to highlight the needs of your security program, prompting executive management to recognize and champion the necessity for a solution.

⟿ Misfire

- Not being ready with the project plan and purchase orders when executive management ascertains an issue and expects you to take immediate action to correct it.

⧉ We are all busy, but are we effective?

It was my first week on the "floor" of the global SOC as the technical CISO. I flew in from out of town for leadership meetings and wanted to meet the staff and see how things were for the folks on the floor. I arrived at the SOC unannounced in the middle of the night. Being a new face and wearing a suit jacket, eyes turned, and everyone wondered what was up. I quickly introduced myself and asked for a small amount of time to sit with the analysts and shift leaders to get to know them and their world. As expected, the shift leader of the SOC excelled in team operations and staff enablement, demonstrating strong organizational skills and a helpful attitude. I enjoyed our conversation. Afterward, I met with various analysts, including one who performed a Wireshark analysis to identify potential threats and report the findings to clients. I asked about their work process, outcomes, training, and what worked and didn't. While meeting with everyone, I wanted them to know I cared and was there to listen so I could make decisions that brought meaningful changes to help them. For the Wireshark analyst role, I taught them how to perform advanced filters and triage within Wireshark to perform tasks much more quickly and efficiently than before. I offered additional training in the future if the analyst promised to share it with others on the job, to which they agreed. My objective as a servant leader who values his people was off to a great start! The analysts had the same roles and responsibilities regarding the overall queue for incoming alerts and responses. This included looking up IPs, cryptographic hash data related to potential threats, and similar components to identify possible threats. They correlated and enriched the threat intelligence by prioritizing, triaging, and reporting their work. Sadly, every analyst had different techniques, tactics, and procedures (TTPs) to obtain these outcomes. One analyst had about 50 windows open in a browser to find "common" lookups performed on the job. Based on the available information, others would have a bookmarked list of a few threat-hunting sites. I inquired about a core set of tools and automation and got various answers that implied it was poorly understood or managed. The analysts did their best with what they were handed or figured out independently. I later found this was the same for all shifts (day, swing, and night).

Sharpshooter perspectives

SSP1: The author's approach was a great way to know the organization's people, processes, and technology. Initially, I found it strange that all the analysts had identical roles and responsibilities, which led me to consider how I would prefer the SOC to be structured. Add well-defined playbooks and consistent tools for the Level 1 SOC Analysts to triage all the alerts. Built-in triggers indicate an anomaly has been discovered that needs to be investigated by a more advanced analyst. When an anomaly is turned over to a more advanced analyst, I expect to see a less structured approach with the expectation that the more advanced analyst has the investigative skills and experience to thoroughly analyze an anomaly without being bound by overly detailed procedures and tools.

SSP2: I am interested in seeing whether the second half of the story gets into the communication between analysts. This model makes sense for a new team working together to become a functioning team. Right now, it sounds like each analyst acts as an individual silo. If management spends the time pulling analysts together or having specific analysts tag team on an incident, the time will be well spent. Afterward, they can develop standard operating procedures (SOPs) and work to automate particular incidents. I am interested in how this will be solved.

Target grouping

The SOC was an operational train wreck, with most team members busy but inefficient and ineffective. The poor performance was easily observable and measured by a lack of customer satisfaction, inconsistent outcomes, high staff turnover, and operational challenges. The managed security services provider (MSSP) culture was to throw people at the problem, churn through staff, and accept high turnover rates while blaming it on click fatigue. Upset clients were managed by offering discounts during the renewal process.

There was a severe lack of leadership, and nobody knew where to focus. Some believed that the technology needed to be changed, which was true. Even more concerning was when I discovered that the development team had been lying, and only IP data was being parsed at a rate of only 7%. Processing IP data was almost useless in SIEM/SOAR and didn't help move analysts toward consistent outcomes on the floor. Some team leads suggested the challenges existed in how "tuning" occurred when onboarding client log sources, including what data was provided, what was actionable, and how the SOC consumed and interacted with client data. The challenge was to achieve the right balance of data and meta-data to support contracts and SLAs. The data needed significant improvements and became a management focus because it was a high client touchpoint. Still, it was not the root cause of the operational challenges. As one Sharpshooter pointed out, functions on the floor were somewhat "flat," so everyone grabbed random tickets and performed whatever functions they could on the tickets. There was little to no maturation of standard operating procedures, TTPs or layers of analytical tradecraft in operations. There were layers of analysts tied to layers of priority, but this didn't do much for efficiency. "Flat Ops" resulted in a daily drowning as many tickets rolled in. In getting to know my team, it became evident that there were many office politics, including gossip and discussions about other people's lives.

Discussing people's dating lives shouldn't be part of everyday operations. This mentally unsafe and toxic culture was a direct reflection of the leadership who lacked character, well-defined expectations, and maturity. The environment wasn't a place for teaming, trust, and mentoring. Lastly, and most importantly, the leadership and professional growth model was what I call "Chief Doers," which is doomed to failure. "Chief Doers" are analysts who are good at following instructions and are the "go-to" staff members who get promoted to management. They may not want to manage people, but they give it a shot since they like the paycheck, responsibility, authority, and autonomy. They may hate the paperwork and leadership components that come with it and may shy away from the politics required to partner with others. They likely have little to no soft skills or leadership training, just "Chief Doer" and "I'm the smartest" or "hardest working" attitude on their chest as they show up to work each day. My focus was coaching the SOC leadership, building a positive culture, and improving the processes.

⊕ Sharpshooter Tip

- Listening is the most important thing you can do as a new CISO. Listen top-down and bottom-up. Meet with each team member and ask for their perspective on the situation, and assume they're telling you the truth. Follow up on your initial findings to discover more and solidify conclusions and theories. Identify core areas of needs and next steps as you lead your organization towards maturity gains and positive outcomes.

Misfires

- It is acceptable to promote "Chief Doers" who have done well by doing so, but you must invest in training them for their new leadership role. They must have a mentor who leads their professional leadership growth and soft skills training so they can be successful. Walk this journey carefully with them to ensure management is a good fit. If not, don't hesitate to move them out of management quickly and into a role they will excel in, like Chief Engineer.
- If your team has lots of busy work and inefficient operations, you may be dealing with the symptoms rather than the root cause. Look closely at the situation to identify all the variables. In this case, technology, process, and culture were all symptoms, with poor leadership being the root cause. You'll never hit the mark if you only treat the symptoms rather than the root cause.

NAVIGATING POLITICS AND PERSONALITIES

One of the things you need to learn early on in your role as a CISO is how to deal with difficult personalities in highly political situations. Very often, you will not be able to dictate action; instead, you will have to use other influence techniques to achieve the objectives for your program, the organization, and yourself. Cultivating this skill can be the difference between growing a successful career or looking for a new job.

Executive war games

I had been at an organization in a vCISO capacity, in charge of redesigning and architecting the security program to take the system to the next maturity level and standardize on industry frameworks. I was finishing the engagement and had been deep in every aspect of the information security program's technical, administrative, policy, and compliance aspects for the last 18 months. At this point, I was the most knowledgeable person on all aspects of the organization's information security program and was the acting CISO. At the same time, they transitioned out the director who had previously overseen the capability.

A recently hired senior executive asked to meet with the CISO as they wanted to receive an overview and status update of the security program. The IT director was the executive over security but wasn't the expert on the program, so he asked me to meet with the new executive. I was happy to oblige, collected the relevant information, and headed to meet them.

Upon arriving, I found the individual seated at the far end of a large rectangular table as if they were a judge holding court. The chair opposite them was about eight feet away, given the table's orientation and where they chose to sit. I introduced myself, walked over, shook their hand, and handed them a folder containing copies of the necessary documentation. Then, I returned to the chair and sat down. After a few minutes of looking at the file, the individual closed the folder and asked me what my title was in the organization. I replied with my role as a consulting resource, but before I could even finish that sentence, they interrupted me, slammed the folder shut, and stated that they couldn't talk with someone so "low on the food chain" and that it would be a waste of their time speaking to me any further. They then asked me to leave and informed me that they would take this lack of respect for their time with the CEO and expected to speak only with their leadership peers on the topic going forward. I was shocked but didn't want to make a scene, so I stood and left the room.

Sharpshooter perspectives

SSP1: What an A-hole! You did the right thing by being respectful and not reacting negatively to their demeaning manner. I hope the story of their actions makes it up to the most senior levels of management and that they will implement the no-A-hole policy. People who act so entitled hurt the operations of the organization. It reminds me of when I first met the CIO of a company acquiring us. The secretary said, "Can you take this call for me." The CIO on the phone wanted to know why there wasn't a limo at the hotel to take him to the office. I said, "Sir, there is no limo because the hotel is directly across the parking lot from the office and is less than a 2-minute walk" That didn't go so well. He still expected a limo. That evening, I witnessed him treating the waitress, who was doing everything she could do, like dirt. At that point, I classified this CIO as an A-hole. That was 10 years ago. Last month, I met a person planning to grant this CIO a prestigious award and told him about the CIO's terrible character. I hope he'll investigate further before offering the award and discover the CIO is still an A-hole.

SSP2: This says a lot about upper management and who they hired for this role. Because the previous CISO was an IT Director, it is suspicious that the previous CISO would also be at the level of the peer. One of the hardest lessons to impart to a deputy CISO is that influence does not directly descend from the title. Instead, influence needs to be cultivated. Most people believe others will listen to them only if they have the CISO title. However, it's usually the other way around: you work with people, they trust your work, and you increase your influence. Over time, you'll become a more trusted colleague.

Executives who don't build trust are shunned and only last as long as their boss does. Those with their own influence are trusted, survive shifts to new management, and excel.

SSP3: A culture of toxicity and power plays is unacceptable, irrespective of your position of authority. When working with difficult people, I always try to default to giving them the benefit of the doubt and understanding where they are coming from and why. Answering the "Why" someone behaves the way they do is critical in truly understanding them. My goal isn't to position them as an adversary but to truly understand what makes them tick. Once I understand their behavior, means, and motives, I can devise a strategy to approach that person best and cultivate a relationship. Some company cultures accept poor leadership and toxic behaviors, as seen in this case, which is unbelievable (and I thought I had some lousy war stories!). Your professionalism and measured response reveal critical elements of any great leader: emotional intelligence and self-awareness.

Target grouping

Dealing with an egotistical senior executive in a highly political organization can be challenging. You can come prepared with the correct information, but certain people will still believe you're wrong. Learning what to do in these situations is a critical executive management skill. Here are some tips to navigate this effectively:

- Understand Their Goals and Motivations: Gain insight into what drives the senior executive. Understanding their personal and professional objectives can help you align your interactions and proposals to appeal to their interests.
- Maintain Professionalism and Emotional Intelligence: Always remain professional in your interactions. Utilize emotional intelligence to manage your reactions and understand their perspective. This approach can help de-escalate tense situations and foster a more productive working relationship.
- Communicate Effectively and Assertively: Clear and assertive communication is critical. Articulate your ideas and concerns confidently, but be open to feedback. Ensure you listen actively to their concerns and respond thoughtfully rather than reacting defensively.
- Document Interactions and Decisions: Keeping records of critical decisions and discussions is essential in a highly political environment. The records will help maintain clarity and protect you in case of misunderstandings or disputes.

Remember, the goal is not to "win" against the executive but to find a way to work effectively within the given environment. These strategies can help you manage relationships and further your professional objectives in a challenging workplace.

In this story's case, I was astonished at the level of entitlement and a little amused as I knew I was the singular resource they needed to talk to, even if I wasn't at their perceived level of importance. It turns out that it was less about me as the primary security resource and more about the expectations set by the executive. Yes, they could have handled it better, and no, they didn't have the best emotional intelligence skills in the world. I discovered the situation was more nuanced than at first glance. I made the most of the challenging situation by maintaining my composure, keeping detailed documentation, and seeking to understand the underlying details. Over time, I built a productive working relationship with the executive. Not everyone is friendly, competent, or even suitable for the role they are currently in. Still, none of that matters, as you only get to choose how you show up in the situations you find yourself in, not the situations themselves. Building relationships with difficult people is one of the most beneficial and valuable skills an executive can develop to boost their career. As a CISO, it is even rarer and more powerful. Remember, you don't have to show up for every fight you are invited to, and most times, the best way to win is to never fight in the first place.

⊕ **Sharpshooter Tip**

- When interacting at the executive level, it is often about you and assuming there will always be a level of information asymmetry you need to bring to every situation. Your goals are not the goals of everyone else. You need to find where there is common ground and operate accordingly.

⌐╍╍⃗ **Misfires**

- Not researching the individual and their motivations. Some research could have avoided the drama and wasted time. If you don't know the context and objectives of the meeting, be cautious or postpone until you do.
- Generalizing a single interaction as a universal judgment on all future interactions. It takes effort, but don't allow one negative interaction to dictate how you approach all future interactions. You never know when you will need an ally, and burned bridges are difficult to rebuild.

SUMMARY

As this chapter on leadership and governance unfolded, it became evident that the role of the CISO is akin to that of a seasoned marksman navigating intricate terrains. Adeptly managing the multifaceted responsibilities of this role requires not only a profound understanding of technology but also a broad understanding of business processes and the strategic capacity to implement controls without disrupting business operations. The chapters delved into the core priorities of leading people and the organization, emphasizing the indispensable nature of perseverance, creative thinking, and adaptability. The importance of instituting a company-wide risk management program, operational risk management, and awareness initiatives was underscored as fundamental to fostering a robust security culture. The significance of crafting and obtaining policy approval was expounded upon, highlighting the necessity for leadership at the highest echelons of management to endorse and champion cybersecurity policies. Furthermore, the chapter delved into influencing with and without authority, measuring security program performance, and the critical role of playbooks, preparedness exercises, leadership, and board awareness. As CISOs navigate the complex landscape of cybersecurity, this comprehensive guidance and practical advice aim to empower them to manage and excel in their pivotal roles as leaders safeguarding the systems and data that make their companies prosper.

Chapter 3

Structured for success

Any CISO who wants to be effective must be "Structured for Success" with a plan that includes short – and long-term roadmap items. This plan should integrate defendable global framework approaches and layers of compliance, regulatory, insurance, and other requirements into its strategy. Balance is critical in how the CISO creates, manages, and adapts their "Structured for Success" plan. It must be coupled with strong leadership alignment, effective delegation, and unification, enabling all leaders to adopt and guide their teams toward successful outcomes.

REGULATORY, COMPLIANCE, AND INSURANCE REQUIREMENTS

The business of being a CISO has changed dramatically in the 10 years prior to writing this book in 2024. Major regulatory, compliance, and insurance changes have affected the CISO, senior leadership, and board of directors.

In 2016, the General Data Protection Regulation (GDPR) changed the global landscape of how organizations manage privacy and human rights. While designed to protect the rights of people in the European Union, those individuals work for international organizations, attend conferences abroad, and so on, with questions and implications on how their information is handled. Furthermore, it set the stage for general privacy and protection rights and permissions, outlining the shared responsibilities between organizations and individuals. It has also attracted significant global attention and real accountability through financial fines and consequences for violations.

GDPR is just one example of how the global landscape has dramatically changed over the past 10 years. A host of cybersecurity frameworks have been introduced or updated, such as the US DoD's requirement for contractors called the Cybersecurity Maturity Model Certification (CMMC), the California Privacy Protection Agency (CPPA), and the California Privacy Rights Act (CPRA). Additionally, the global threatscape has intensified, with cybercriminals using ransomware to extort massive payouts from large organizations. These attacks have also changed how insurance will and will not

DOI: 10.1201/9781032720500-4

pay out for security incidents. All these changes are having enormous implications for CISOs!

CISOs must now carefully plan and navigate multiple layers of regulatory, compliance, and insurance considerations to ensure that they make the right decisions before disaster strikes to ensure cyber resiliency. For example, will you as an organization consider a payout if ransomware strikes? Could you quickly pay a ransomware payout in Bitcoin? If you do, would it violate any terms and conditions in your insurance coverage, potentially preventing you from receiving payouts to cover your losses?

Planning for multiple scenarios and developing runtime playbooks for incident response teams, legal, third parties, and others require due diligence, which a CISO must consider and exercise to be successful.

⮂ We Don't have any compliance requirements

When considering a third-party software-as-a-service (SaaS) provider, we reviewed vendor compliance and security standards to identify risks before accepting and onboarding a relationship with a third-party provider. As with most startups, it was a smaller shop with a few dozen staff, mostly younger, who all had a script and standard pitch for selling the service – but I needed to know if they were secure and compliant. When I asked the initial sales lead staff about security and compliance, they stated they didn't know the answer. When I asked sales staff about security and compliance, they stated they didn't know the answers but would find out and let me know. My questions were basic such as, "Are you ISO or NIST compliant or have any security certifications or attestations you can share with me."

After processing the vendor, with a few emails and meetings over a few weeks, I finally got a meeting with their Director of Security. When I spoke with this individual over online conferencing, I would ask about architecture, security controls, design, testing, and more. I repeatedly received answers about their Google Cloud hosting solution and its capabilities. I never heard things about how they, as a company, took responsibility, ensured security outcomes, or managed security within their operations, their app that we would be using with our clients, or their third-party relationships – always what Google could do and what capabilities they had enabled in the cloud.

After asking for certifications, I was told by the Director of Security that they were not required to have certifications, yet they were in the medical field. I understood they were based overseas, so they were potentially not required to meet specific health regulations and requirements in their own country. Still, I asked how they comply with global standards when selling a SaaS solution to international clients with needs that must be met, such as HIPAA (health insurance) and GDPR (privacy). They stated they had no such requirements but could consider that on their roadmap. I then asked if their app had ever been penetration tested and was told that it had not been, but they were working on it. A month later, we received an unexpected update when we were moving away from considering this vendor. This penetration test was completed earlier in the year, revealing massive security gaps and issues, with most security issues unresolved upon retesting. I noted the dates of the test and retest, which had taken place before my previous conversation with the Director of Security.

Sharpshooter perspectives (SSP)

SSP1: This scenario is very common for SaaS suppliers. They usually haven't been in business long, and their business model is to keep prices very cheap by forgoing common security and business practices and electing to limit their risks in the terms and conditions. In this scenario, the SaaS vendor has no clear understanding of which security controls are provided by the Google Cloud Platform and which ones are their own responsibility. Through stack mapping, all major host providers, Google, Azure, and Amazon, define their security controls. The SaaS provider is responsible for the rest. Without these controls, this SaaS provider cannot be HIPAA compliant, and the author did the right thing by walking away and not subjecting the company to the associated risk. Since this is HIPAA-related, I'd also add a Business Associate Agreement (BAA) review to see what the SaaS provider agrees to protect for the covered entity.

SSP2: I am curious about the title of the security lead. Was the person a security architect, or were they a manager of IT that the company asked to take on the security role part-time (with no training)? The refrain of Google, Amazon, or Microsoft handling our security is too common, and a security person knows that it matters how one configures Google, Amazon, or Microsoft. The other common refrain is "We are used by [insert name] of a big company," meaning they must be secure if a big name uses them. That is why we ask for outside confirmation, like the certifications. From the story, I am guessing that the security lead is more of a salesperson because not knowing that a penetration test had been performed, even if the results were awful, tells if the individual is aware of the product's security.

Target grouping

The SaaS company's Director of Security was not qualified for such a role and was not truthful about onboarding third-party services. Most security answers reflected security capabilities from their hosted cloud services provider, not provider-owned and managed security measures (big difference).

It was clear from the discussion around penetration testing that when the original ask occurred for attestations and evidence, when the Director of Security clearly stated they had not performed any such tests to date, that he was lying, as later when they shared a report to rescue a possible sales lead that was falling away. The penetration test showed that issues identified months prior had not been resolved, even after retesting. This revealed clearly that the Director of Security had performed penetration testing and could not resolve critical and high vulnerabilities discovered in production that were relatively easy to remediate. He then lied about having performed any penetration testing to date because he didn't want to admit to that ugly truth.

⊕ Sharpshooter Tips

- In the cloud, you have a shared responsibility and attack surface. If you're not personally up to speed on these complexities, get someone in your wheelhouse who is, as the advent of cloud, SaaS, and other solutions is gaining speed here in 2024. Your shared risk and responsibilities in the cloud and how to navigate that from a legal and attestations perspective are critical. Doing that upfront when you have all the power in contracts, attestations, and negotiations is essential!
- SaaS salespeople who may have a "security" title or role may come off as security people when, in fact, they are salespeople. Do not be easily convinced or hear what they want you to hear. Scrutinize their experience, certifications, and background to address your security needs as you investigate security requirements for your shared responsibility.

⌁ Misfires

- Never trust a SaaS provider outright. They are often young startups who may cut corners, including using software full of critical vulnerabilities and lacking security protocols and best practices, while telling you that their data is encrypted (referring to standard cloud volume encryption), which isn't doing much, if anything, to protect your data. Ronald Regan was right when he said, "Trust, but verify," where attestations and looking under the hood are required before buying.
- The cloud has a shared security model that requires careful consideration and management from all parties throughout the lifecycle. In SaaS provisioning, the majority of control and responsibility for selecting the SaaS provider is upfront when you have the leverage and cash to require contractual SLAs and terms to meet best practices and other conditions to lower risk. Simply acquiring a service and accepting default conditions from a SaaS provider is always a foolish move as a CISO.

⊇ We've Got a backup, right?

This question floored me in the middle of an incident by an executive of the victimized company. "We've got a backup, right?" My company, which specializes in incident response and remediation, was brought in to identify and remove the threat – ransomware! Ransomware actors changed tactics around 2020 with more targeted attacks in two realms: cheap and fast for individuals and whaling-like targeting of organizations. The bad actors also used new tactics, like exfiltration, to force huge payouts from big organizations with deep pockets (from an actor's perspective). The medium-sized organization that hired us was in the middle of the incident in which their files had been encrypted. They had no visibility of what exfiltration may or may not have occurred, and they were in a war room with their consults, discussing what to do next. Considering their options, they didn't know the quality of the backup or if they should consider paying the ransom to get their data back and become operational post-incident.

Sharpshooter perspectives

SSP1: Well, if they don't know if they have a good backup, chances are they have no plan on how and where to restore the backup. With reservation, I'd start evaluating how much the ransom is and how I'd pay it. While doing that, I am also looking for any backups and, if found, testing them to see if they are good. I'm also looking at the "integrity" of the ransomware group and the level of assurance they will not leak any exfiltrated data if I pay the ransom. Even if I pay the ransom and if, in good faith, the hacker provides the encryption key, there is no guarantee that the files can be decrypted and services are restored. There is always the potential that the files have been encrypted by more than one hacker organization, and if so, multiple keys will be needed, and they must be applied in the correct order.

SSP2: I wonder if the executives are well versed in the tactics and asking, "Did the attackers corrupt our backups?" That is a very legitimate question. However, if the questions are, "Do we have backups?" or "Can we restore from our backups?" then, it sounds like business continuity and disaster recovery (BCDR) has been neglected. If it is the latter set of questions, then it is likely time to ask a negotiator for the ransom and ask someone in IT to test if they can restore a backup.

Target grouping

Fortunately, the company did not suffer a massive ransomware breach or multiple forms of extortion. They dodged a bullet!

The incident response scene was a true crisis because leadership had not prepared. There's so much wrong from an incident response (IR) perspective. I'll start with a high-level bulleted list to summarize the critical points!

- IR planning with play/runbooks for common threats and risks, including ransomware. Everyone involved in incident response should be well trained and know how to follow the play/runbooks in case an incident occurs, such as the ones above.
- Backup and restoration are a life cycle of operations that should be ongoing for all critical files and tested regularly, especially considering

the prevalence of ransomware in 2024. This process is increasingly complex with hybrid infrastructure. You'll likely have diverse backup and restoration functions and components in your overall plan that you test and validate monthly. Don't wait for an incident to discover your backups and restoration processes aren't working as expected!

- Escalation planning must exist with pre-established relationships, including legal and local authorities. These relationships are becoming more critical, especially if you're a publicly traded company in the US dealing with a material event and/or incurring losses or damages of $1 million or more.
- Ransomware readiness is its own beast – are you ready? What conditions and criteria do you have to follow to meet legal, insurance, and other requirements to consider a payout? Is it even legal in your area, because in some places it's not? If you intend to pay as a contingency, can you? Do you even know what bitcoin is, how to monetize millions in bitcoin for a bad actor, or how you may want to execute that transaction in such an event? Effective planning, thorough preparation, and creating accounts are necessary for speed and readiness if you're hit with ransomware.
- Data integrity matters. What controls do you have in place to ensure that, if you're performing recovery following an incident or potentially receiving data back from bad actors, you don't unknowingly end up with compromised, injected, or weaponized data? Golden images, approved software, hash lists, and similar solutions are required to assure data integrity.

⊕ Sharpshooter Tips

- Identify the organization's top threats and risks and ensure you're prepared to identify, respond to, and proactively reduce risk against each. As the CISO, this is one of your essential responsibilities. Ensure your highest-value assets have the best security measures to protect you from the most significant risks.
- Backup and recovery are often shared responsibilities between IT and security, requiring complete tests to ensure operational effectiveness and readiness. A BCDR program requires business units to identify roles and responsibilities (RACI) and regularly validate that essential backup and recovery functions are operational. Don't wait for a significant security incident like ransomware to discover your BCDR isn't functional as expected. You must commit to training your team, operationalizing, and testing the BCDR to confirm you're ready for a disaster.

⌁ Misfire

- "If you fail to plan, you are planning to fail" – Benjamin Franklin

⛁ We Don't have any participants from the EU

As I helped a group of nonprofit volunteers plan a regional cybersecurity conference, the discussion about privacy came up about the GDPR and how it might apply to registered participants. GDPR was a hot topic in the news regarding the privacy rights of individuals in the European Union. The volunteer was implying by the question that GDPR did not apply to us because our conference was in the US, and we didn't know of any European participants in the past. We needed to decide what was legal and wise to protect the data of participants registering for the conference. We also needed to determine how to manage and archive their data and the legality of sharing their information with sponsors who paid for it as part of their package.

Sharpshooter perspectives

SSP1: GDPR is irrelevant here since this conference is in the United States, and you are not conducting business in the EU. What is relevant is that you protect participant data per the applicable state privacy laws. Start by telling participants what data you collect, how their data will be used, how their data will not be used, and then do what you say. Strongly consider an "opt-in" clause where participants choose whether their PII is shared with the vendors. If they don't "opt-in," none of the vendors get the participant's PII. I'd also include a statement that they authorize the conference to notify them of the same conference in future years by registering.

SSP2: Ironically, the data protection question is being argued at a cybersecurity conference. It reminds me of a vice president who argued that we did not need to protect people's credit card data because most people would only be liable for $50. Besides, if their data were stolen, we shouldn't notify them because we wouldn't know if it would be used. GDPR can be invoked for anyone who is an EU citizen. As the prior Sharpshooter stated, the better course of action is to consider an "opt-in" clause. Define the rules for what type of data you collect and put in rules for how it is used.

Target grouping

Just as seen in our Sharpshooters Perspectives, there was a varied understanding of the implications of GDPR and how to navigate the privacy requirements of conference registrations wisely. The accurate answer is that any European Union (EU) citizen is protected under GDPR law for any company doing business in the EU. Suppose an EU citizen attends a conference or signs up and discloses their data during registration. In that case, the organization is responsible for and accountable for securing and protecting the individual's personal information. However, because the conference is in the US and the company is not doing business in the EU, it is likely not subject to GDPR (no legal jurisdiction). Regardless, the organization may be subject to local, state, federal, national, and other laws and regulations similar to GDPR. Forrest Gump was right when he said, "Life is like a box of chocolates; you never know what you're gonna get." Consult an attorney for legal advice specific to your business and location. CISOs would be wise to universalize company privacy to GDPR-like standards for unification and ease of global administration, especially for business units with financial and regulatory requirements.

⊕ Sharpshooter Tip

- Universal privacy wisdom is relatively easy to achieve. GDPR attempts to regulate what should be common sense in some regards (this could be debated) in collecting, managing, removing, and performing lifecycle management of data willfully disclosed by someone. We are compliant if we afford these privacy rights to everyone, ensuring we've met local, state, federal, national, and GDPR laws. Usually, this rolls up to the point that if you're performing GDPR-compliant practices, you probably comply with all other similar privacy laws. Consult with your legal counsel to ensure compliance.

⌐͞ Misfire

- Never make decisions based upon the "opinions" of others, who are often misinformed. If you have doubts, consult with legal counsel to understand the law and determine how privacy, contracts, and legal matters apply to you in various situations.

CYBERSECURITY FRAMEWORKS

As a CISO, have you ever faced challenges aligning priorities with your staff, advancing specific overarching goals as a team, and achieving unification? Cybersecurity frameworks not only provide the required structure but can also remove the people and politics angle, where you focus on the process instead of personalities! This structure is very beneficial when dealing with turf wars, egos, and cultural challenges. Frameworks can also help you meet specific standards that qualify you for insurance and, in some cases, lower premiums.

Embracing cybersecurity frameworks can begin with understanding "WHAT" should be included in a SecOps program, such as the NIST Cybersecurity Framework (NIST CSF) with approximately 100 controls. Start by learning the process and conducting a self-assessment, then have a third-party audit to ensure integrity, further refine your approach, and strengthen your team's capabilities. Once the initial framework is operational, consider upgrading to the more extensive and comprehensive NIST 800-53 or a similar cybersecurity framework. This upgraded framework should not only outline the necessary components of the program but also detail how these components are achieved. Additionally, it should assess the maturity of operations through process and documentation, measured against the international Capability Maturity Model Integration (CMMI) model. This will allow for the baselining and maturation of operations based on business objectives, key performance indicators (KPIs), and various metrics and goals.

⅀ I need a **NIST CSF** assessment completed in dozens of countries in 2 months or less

When running a global cybersecurity framework consulting practice for F100, a client onboarded and asked for a baseline of operations to the NIST CSF within 2 months. Completing an extensive assessment within a month is feasible if the client can ensure timely participation and flexible availability of interviewees, prompt provision of requested documents, and regular meetings. This will allow sufficient time for data collection, interviews, analysis, and reporting. The client then stated their operations took place in dozens of countries – and yes, they wanted it all completed in 60 days. The desired outcome was establishing baseline operations, securing management's commitment to adopting a framework-driven approach, and developing a roadmap for maturing SecOps.

Sharpshooter perspectives

SSP1: My first question would be, "Does the CSF apply equally to all operations, or will there be a difference in the framework tiers: Partial, Risk Informed, Repeatable, or Adaptive?" The expected framework tiers must be established before a gap assessment can be completed. Second, I would ask if specific business units are at higher risk than others and suggest increasing emphasis on completing those assessments first. Third, I'd approach this as a remote assignment for most, only having boots on the ground for the highest-risk areas. Fourth, I'd ascertain if there are F100 employees with cybersecurity knowledge and foreign language skills and then map them to the number of assessments that must be performed in that language to determine if enough resources are available. If there is a shortage, I would request the client provide translators. From there, I'd create a quick proposal, including an implementation plan and costs, to understand their interest in completing this in 60 days.

SSP2: I would determine how invested the company is in this project. If this directive comes from the CEO, emphasizing to all units the priority of completing this baseline and indicating that penalties will be imposed on teams that do not participate, the project becomes possible. In this scenario, a project plan with allocated resources can be developed. However, if this initiative is not driven from the top, it will unlikely be completed within 60 days.

Target grouping

The project was aggressively managed across multiple geographic regions with various languages, and operational challenges unique to each area. We completed it within 60 days and followed up with additional work specific to alignment, consumption, and next steps. To no surprise to the author, the client became very difficult, critical, and divisive, bringing the project to a crawl. The stalled long-term adoption of a framework-driven practice quickly became evident as we better understood the organization's negative culture and top-down leadership.

One Sharpshooter accurately discussed starting in areas of highest risk or areas of priority. As a CISO, why would you begin with a challenging project with such an aggressive timeline? Was this to assert authority and try to make your staff care? Does that build motivation and team building and get them behind a framework-driven operation? No. It will likely make them resent a big push, resulting in nothing but busy work and "ammo" for the big boss to use against others and an erosion of trust. Trust is destroyed, and leadership is weaponized within the framework if utilized in this way.

Turning back to what made the project work, here are a few takeaways:

- Reiterating that the client must have high commitment and availability of resources during such a tight window. If someone was on vacation for 2 weeks, how can we stage interviews, follow-up inspection of documents, and other activities related to the assessment? Daily operations were established to identify risks and manage them accordingly.
- High-trust relationships, which resulted from daily operations, were essential. We worked hard to be dependable team players – responsible, professional, and agile – all striving for successful outcomes. We also connected personally, which paid off in spades when stressful situations occurred later. Trusted insiders shared with us and partnered toward outcomes to help us turn the corner with the toxic leadership above.
- Organized roles and responsibilities into highly detailed groups, broken into geographical language and technical components and mapped to dependencies. This information resulted in successfully scheduling critical interviews and other activities required for the assessment.
- We focused on establishing a culture of the CSF as a baseline of truth, "good, bad, or ugly," with no shame in whatever the outcome so that we could move toward maturity. This positive culture was later undermined by toxic senior leadership with weaponized motives. We gained the respect and trust of leaders during the assessment, but senior leaders destroyed them during the reporting and consumption phase of the assessment.
- NIST CSF is just the beginning; it establishes "WHAT" is to be in your SecOps program without the more significant "HOW" measured in NIST 800-53 and other frameworks. It is an essential foundation for

any organization committed to adopting an iterative, framework-driven approach to systematically structuring operations, reducing risk, and ensuring compliance in SecOps. This engagement resulted in a solid CSF baseline, but the client didn't perform any follow-up or adoption toward CSF or NIST 800-53.

⊕ Sharpshooter Tips

- Listen to leaders to hear what their direct reports are saying and not saying. This information will tell you all you need to know about an organization's culture and vibe and why it functions as it does or does not.
- Design a framework-driven SecOps program with evidence at the center for validation and audit support. A centralized database with artifacts explicitly mapped to controls, proving the control is implemented and followed, is evidence management needs, along with auditors, to prove you're doing what you said you're doing in operations. Suppose you have policies and procedures and can evidence this in process and operations. In that case, you're at an industry-standard best practice, CMMI Level 3, which is fantastic compared to most, which tend to be more ad-hoc or people-driven than process-driven.

⌐⌐ Misfires

- It's not enough to simply perform the NIST CSF or have a third party assess you. How you choose to wisely promote a safe culture of transparency to adopt the truth and maturity of SecOps in operations is critical for future success. The CISO is essential in prioritizing a framework-driven mindset and maturity in SecOps.
- Don't be the CISO looking for "ammo" to beat up his staff. This mindset is demotivating and shaming at best. Nobody wants to work for a person like that. Be the best you can be and help others do the same.

COMMUNICATING CYBER RISK

Get all your leaders into a room and ask them what cyber risk is, how you measure it, and what the difference is between low, medium, and high cyber risk. Then, ask them what the low, medium, and high-value assets are and how they are protected. You will likely get answers all over the board! In our industry, we toss the word "risk" around as everyone understands it, but we don't. For example, in the world of vulnerability risk management, the Common Vulnerability Scoring System (CVSS) is used to provide a severity score to a vulnerability on a ten-point scale, where, say, a score of 8 is pretty high, but that is NOT risk, it's the severity of the raw impact of that vulnerability. It does not represent the risk to a specific organization or the likelihood of exploitation.

Cyber risk is made up of two core elements: impact and likelihood. How you socialize risk management and, more importantly, unify your leadership team and your overall culture and company is a challenge! It would help if you intentionally devised messaging strategies to communicate risk in everything you do – daily, weekly, monthly, and annually. Design all your programs around risk. Implement matrices, policies, and procedures, and ensure everything you do is based on risk-based decisions, data, outputs, and reporting. Then, reinforce this reality and vision while moving forward. Ensure your leaders are all doing the same, in the same way, helping you all mature operations and risk-based outcomes. This cultural change and shift takes around 2 years in most organizations, so plan for the long term.

⮂ Perceived versus actual risk

After handling an incident for a new client, we onboarded as a consulting agency to support the maturation of their SecOps, including cultural and leadership changes in the organization. The security team claimed there were many challenges, and in reflection of their incident and other events, they had many security challenges to overcome in operations. Meanwhile, senior leadership and elected officials were seemingly in a "far off land," removed from this reality, thinking everything was working and must be OK. That was the working perception, so we created an offsite workshop to address this alignment and leadership concern. During the offsite, we asked everyone in the room to first-rate how secure operations were, on a scale of 1–10, and to write that down. We then asked everyone in the room to raise their hand if it was below a five or five and above, with revealing results from those who raised their hands in each group!

Sharpshooter perspectives

SSP1: How was the security team positioned within the organization? It sounds like they were far down the chain, so their concerns were filtered through several layers of management. It's possible upper management wasn't hearing the issues within the organization because the challenges weren't being filtered. If that is the case, it is time to restructure the organization.

Another possibility is that the security team has trouble articulating risk to upper management and has become known as an organization that cries out that everything is a risk, stalling every project. The exercise will likely bubble this to the surface. If that is the case, it will be a project to train the security team to better interact with upper management and move forward after this incident.

SSP2: Words mean things. When there isn't a common understanding of the meaning of things, creating a shared understanding of the effectiveness of work efforts and the programs they support is extremely difficult. That gap in understanding can devastate an organization's accurate risk posture and create a situation of security theater that does no one any good.

Target grouping

The client was part of a state agency lead, different from standard commercial politics, where there was a notable gap between IT and SecOps and management. Sharpshooters see the gap clearly where "la la land" is real for IT, frustrated by being ignored in budget, priorities, and voice. The offsite exercise was an opportunity for a small workshop activity to create awareness and alignment where unity could be achieved, which it did with outstanding results. There was a huge gap between all the IT staff rating security as very low and senior management rating it as very high – how revealing! We let it sink in and then challenged senior management as to why they think it's so secure when the block and tackle people know it is not!

Cultural change takes a minimum of 2 years. Two years into the engagement, we moved from coming in as a company supporting an incident to discovering multiple incidents, helping SecOps develop a comprehensive strategy, and performing baseline framework assessments and advisory services, including the exercise in this story. The SecOps team experienced a notable cultural change and traction with management. Things were moving swimmingly toward adopting a full NIST 800-53 driven SecOps program, run/playbooks for IR readiness against top threats for the organization, and increased training for the IT team until, staffing changes occurred a few years later. New leadership was selected, and the contract quickly dissolved; IT became silent, and the status quo returned.

⊕ Sharpshooter Tips

- Cultural change must be intentional, based on what you're messaging, how you build trust, and who and how you influence others.
- Cultures and SecOps programs that are immature need help in understanding terms, alignment, and unity, with lots of grace and encouragement along the way. Start by defining terms in simple and easy-to-understand words. Risk is easily calculated in terms of two variables: likelihood and impact. Then, a simple workshop should be held to drive accountability and involvement from stakeholders by asking them what high-value assets are at risk for loss, injury, or adverse circumstances. Ask what types of losses might be incurred, such as operational, reputational, and financial, as you guide them through a new way of thinking and doing to become a risk management-driven operation.

⌁ Misfire

- Don't assume that operations in a state, local, and education (SLED) environment are the same as those in a commercial environment; they are dramatically different. SLED is usually based on more personal relationships and bureaucracy, subject to local politics that are difficult to challenge without long-term leadership committed and capable of the fight required to overcome entrenched SLED politics.

₴ This is the likelihood – not

While I was leading a global threat and vulnerability management consulting practice for an F100 organization, an organization asked for assistance in creating more efficient and prioritized operations to handle a growing backlog problem in production. To our surprise, we discovered over 50,000 high and critical vulnerabilities in production! Risk management proved severely limited in this environment, with leadership "accepting" massive risk instead of allocating a budget to resolve what most would consider security essentials.

It was clear that, over time, a large amount of "tech debt" had accrued, creating fragmented and inefficient operations, such as a patch schedule with nearly two dozen stages. This inefficient process led to discussions around desired KPIs and reduced risk outcomes, such as patching for current "N" state goals within 30 days of Microsoft's monthly patch release and ongoing releases by other software companies. This new process enabled us to redesign the patching schedule around these core priorities to make it more efficient.

Beyond the patching schedule and tech debt, there was a great deal of ambiguity, and the root cause of patching was genuinely unknown. Why is this server still unpatched, even though you attempted to patch it? How do you validate, track, measure, manage, and improve the process?

Perhaps worst of all, it was a culture compelled by the CVSS and how vulnerabilities were scored for base, threat, environmental, and metrics representing the raw impact of a vulnerability. The organization equivocated this to the likelihood of exploitation and, in general terms, the risk of vulnerability. As a result, staff used the CVSS number to boldly state that the possibility of a vulnerability with a CVSS score of 8.5 was very high.

Sharpshooter perspectives

SSP1: A Common Vulnerabilities and Exposure (CVE) score from CVSS is only part of the picture when determining a vulnerability's actual risk to an environment. Countless times, I have received emails from executives that the sky is falling because a particular vulnerability is in the news, only to reassure them in many cases, due to configuration, alternative control measures, and other reasons, that the vulnerability poses little to no risk to their environment. An understanding of the impact of exploiting a vulnerability is missing within this organization. I was in a similar situation when we first started to evaluate the security risk involved with process control systems. No one took security seriously until process control networks were attached to the internet to allow remote monitoring of processes. Connecting the network to the internet completely changed the risk profile since the likelihood of infiltration increased significantly, increasing the risk of compromise. In contrast, before, the likelihood of a compromise from the sneakernet was very low.

SSP2: In 2007, the book *The Pragmatic CSO* by Mike Rothman influenced my thinking on handling vulnerabilities. Mike pointed out that there were unending audits and vulnerabilities, but one of the things you need to do is to prioritize your systems. He recommended asking the other executives what systems they considered essential to run their business. Usually, each executive has a list. It was a quick way to understand what the executives value and how much they value it. Although that methodology is only part of the patching story, it does allow you to focus on how the CISO's reputation is impacted if they don't protect a system that an executive expects you to protect.

What appears to be missing here is a methodology for determining what systems are essential and those that are exposed and then working to patch those first. This methodology begins to solve the risk acceptance problem, as executives will typically fight to protect their system once they understand the actual risk. If not, they just accepted the risk!

Target grouping

This organization's culture of minimizing security persisted for years, resulting in massive technical debt and inaccurate methods of understanding and communicating risk. Even when they started to feel the pressure from a whopping 50,000 high and critical CVE CVSS-rated vulnerabilities in production, they misunderstood the risk and, therefore, massively misunderstood and accepted it. Let me say that again: they had massively misunderstood and accepted the risk without realizing it was the root cause.

We solved challenges in this environment by systematically and affordably addressing issues they could understand, starting with their patch cycles and priorities and then tackling other problems that revealed themselves in their threat and vulnerability management (TVM) lifecycle. However, this does not address the broader issue of risk acceptance or improve how executives understand and communicate risk. It merely resolves a few problems that any experienced consultant could have easily identified and fixed.

⊕ Sharpshooter Tip

- A CISO owns the entire problem, not the symptoms, ensuring they understand the context, the full risk context, and all that is being accepted and mitigated regarding risks and threats.

⤳ Misfire

- Ignoring the issues, listening to the status quo, and being part of the "Ostrich Syndrome" is part of the problem. Don't contribute to dysfunctional risk by accepting a culture of denial or risk acceptance that leads to a mountain of unknown or accepted risks, exponentially increasing the attack surface and unmanaged SecOps.

⏚ impact – not what I expected!

I learned a lesson early on in security when the SQL Slammer worm wreaked global havoc in 2003. A patch was available months before the worm's launch, so there was plenty of time to patch, and patching we did, or so we thought. When the worm hit, it became a better auditor than our own processes. What we thought was patched – or instances of vulnerable versions in embedded solutions we didn't know were on the network – were all revealed in rapid succession as the worm spread like wildfire across the network. SQL Slammer was well-named as one of the fastest "superbugs" to date, with few default security controls in place to limit techniques and tactics leveraged by this file-less threat in 2003! It doubled in size every 8.5 seconds, with over 90% of vulnerable hosts infected within 10 minutes and over 55M scans per second after 3 minutes! The impact of this worm was far greater and unexpected than we could have imagined before the attack.

Sharpshooter perspectives

SSP1: For me, it was Code Red in 2001. I was working at a Fortune 500 company as a Linux server admin. We had hundreds of thousands of devices on the network. Unlike other people, I read my logs. One day, I noticed all these strange things in the logs on my internal servers. After much research, I sent my log information to the security team, who mainly did policy and access control. The Code Red worm had gotten into the company and was rampant. Every day I read logs and sent the information to the security team with the latest infected IIS server in the environment. Like this story, we discovered we had IIS everywhere, such as the IIS server that someone had installed on a kiosk meant for people to check email. The biggest lesson I learned was not about patching or worms, but finding a company that valued what I could do. Although I was sending and tracking the infected IIS servers to the security team, they never responded. Later, one of the server admins took me aside and said that my logs were getting forwarded to individuals within the company to remediate the infection and that it had made a difference. After Code Red, we got hit by Bugbear, Sasser, and SQL Slammer. I tracked a lot of infections for them. Although that company did not value my skills, I got picked up by another firm and left for a successful career. Ironically, my former company was hit by notPetya and was shut down for a while, making the news. The CISO lesson I learned from the early worms was that one of the best defenses is recognizing and cultivating security talent, keeping them engaged in protecting the firm, and letting them know they are valued.

SSP2: I learned three things from one worm during this same timeframe. (1) It takes only a few computers on the network broadcasting to make the network unresponsive. (2) All systems on the network must have up-to-date patches. (3) You must be able to remove any computer from your network.

We were hit with a worm from a computer we did not own but was connected to our network. We had installed a monitoring solution and could see the worm spreading. We stopped it! Not really; it infected 72 systems since those were the only ones susceptible of the 15,000 systems. All the others were up-to-date on patches. We had to remove all 72 infected computers from the network to get the network back into operation. This incident occurred on a Friday when the DOE was on the golf course before cell phones were common. We couldn't reach them, and most of the computers were theirs. I made the executive decision to go into their space and disconnect their infected computers. On Monday morning, I was called on the carpet to explain my actions. I had a sound case of why I did what I did, and DOE management agreed that I did the right thing. But was I authorized? The answer was no, but this incident led to formal authorization and special badges being issued by DOE for people in computer security to decide to shut down, disconnect, or even confiscate computers.

Target grouping

I've learned over the years to plan before a crisis, yet I found myself in a situation where I felt like I had failed. I saw an opportunity to learn. The nature of technology, dependencies, and complexities of patch management is not trivial in 2024, where continual learning is critical. In this incident, the key issue wasn't the worm being fileless or its rapid spread; it was the lack of visibility into the nested dependencies within our TVM practice. How can we identify and address vulnerable nested dependencies or software components that we may not be aware of?

Additionally, how do we ensure that updates are rolled out or that controls and workarounds are implemented across various products to mitigate risk effectively? That's a huge ask in such a nested "Russian Doll" of patching complexity seen in this case. We learned a lot and adapted and improved our TVM programs.

Slammer also reminded us of the basics of incident response (IR) and the value of planning, preparing, and stress testing before a crisis:

- Create a runbook or playbook for the most likely threats and highest risks to specific business assets, prioritizing them according to your risk management plan.
- Regularly review and update incident response plans. Perform tabletop exercises to test the plans and, based on the results, identify and improve IR planning, processes, and awareness.
- Architecturally move to the left of boom by adopting concepts such as Zero-Trust Architecture (ZTA), segmentation, isolation (e.g., software-defined networking), identity access management controls, and other security measures. This proactive approach reduces the blast radius, ensuring that if an incident or intrusion occurs, it is confined to the smallest possible architectural area and set of assets.

- Ensure that, operationally, you can respond as quickly as possible, 24x7x365, because threats can move at light speed.
- Review and follow emergent threats and TTPs to ensure your IR jump bag has the tools and tactics to detect the latest and most significant threats, such as identifying and capturing in-memory-only payloads.

⊕ **Sharpshooter Tips**

- Continual learning is critical as a security professional. Even with excellent planning and leadership, moments of failure and defeat are inevitable. Your war is won by responding to each battle.
- Hire and retain great staff, using every means necessary. People, not technology, make the difference in a great SecOps program.

⌐💨 **Misfires**

- If you don't do the basics well, you'll never have a chance at being effective, let alone a great CISO. Master the foundations of security, like auditing logs, patching, and leadership, and leverage these foundations into transformative leadership programs.
- Never assume employees will choose lifelong learning by default. Human nature requires a "shepherd culture" with a leader who rewards behavior for those who learn and grow.

ᘔ Malware jeopardy

While managing a new cyber threat intelligence (CTI) team, I found that each team member had a different understanding of risk and a unique approach to measuring it. Some team members had a military background, with a focus on human intelligence and traditional operational support and intelligence. However, they were less experienced in cyber security and CTI operations. Other team members came from the cybersecurity field and had college degrees but were still junior and had less experience, training, and understanding of risk. The team had previously been led by an individual focused primarily on geopolitical threats. As a result, he needed to improve the balance of tactical, strategic, and operational intelligence to better align with the specific goals and objectives of the company's business units. Each time staff members generated reports and rated risk, a new metric was communicated in all intelligence reports. Surprisingly, I found significant variations among the staff in how they viewed, scored, and communicated risk. This discrepancy became apparent with a malware threat; some rated the risk as low, others as moderate, and I rated it as high for our organization.

Sharpshooter perspectives

SSP1: Differences of perspective concerning risk severity are common and healthy for a vibrant risk management program. A well-defined, systematic, documented risk process has worked for me to establish successful risk programs. Get everyone in the room together, whether physically or virtually, discuss the risk, hear all perspectives, and then assign the resulting scores. What occurs during this process is the team understands the risk better and has an appreciation for each other's perspective. If there is fundamental disagreement, take a poll of the risk scores, throw out the highest and lowest, average the remaining, and move on. What will happen over time is the scores will be normalized.

Don't make it overly complicated, so everyone participating must continually focus on the process rather than the risk. Keep it simple, but be sure to include a severity rating, how it will impact the business if it occurs, and the likelihood and the chance it will happen in the risk rating. These will also be key areas of focus during risk mitigation. If I lower the impact or the likelihood, I've reduced the risk, which is the goal of any risk program.

SSP2: I encountered a similar problem in that everyone on the team used their own methodology to examine how an attack progressed, while I needed them to use a standard like MITRE or Lockheed Martin's kill chain. The couple of times this happen, I got everyone to standardize on the Lockheed Martin cyber kill chain. The interesting piece is how much the group culture and training influenced the decision. I gave them a couple of standards in one team and asked them to return with a recommendation. It was Lockheed's model because the analysts felt it fit closer to what they were experiencing. In the second case, my team was mainly former military, and they kept asking me to choose for them. I chose Lockheed's model, although I could have chosen either. For both cases, the model led to a more focused vocabulary for defining risk and produced better consistency.

Target grouping

As one Sharpshooter stated, it can be as simple as getting everyone in a room. A small amount of structure to ensure it's safe and has outcomes is all you need. Avoid burdening the staff with overly structured or managed "unity" exercises that may challenge trust instead of championing it. Instead, the author had an honest, nonjudgmental discussion about risk with individuals and the team. This was followed up with documentation of a policy and procedure related to a risk identification matrix and decision-making. Subsequently, a "pilot" workshop was held where everyone gathered to apply this new approach to improve the identification and communication of risk. Most importantly, the workshop was presented as a pilot, with an open invitation for feedback. Adjustments would be welcomed to ensure the process met our needs and achieved our desired outcomes.

Following the matrix exercise, each staff member was required to have a monthly growth goal specific to them. Additionally, they had a goal related to risk management and rating, which would be performed and managed monthly. Each staff member was also paired with a mentor to provide daily and weekly block and tackle support, addressing challenges and questions. This mentorship created a sense of "team" as everyone learned something new, reduced anxiety, and promoted collaboration and relationship-building. Over time, the team's consistent performance led to the development of on-the-job training (OJT) requirements tailored to various roles and levels of analyst maturity. This success also paved the way for promotions and additional OJT programs for new hires and employees joining the team from other business units.

One successful part of the OJT program was allowing new team members to challenge themselves and attempt new tasks without fear. For example, the person could handle a malware risk they were not yet certified on, do their best, and then have their work reviewed by a senior, certified staff member.

The senior member then provided positive feedback and coaching to guide the new team member toward certification. This approach streamlined the process for those with natural aptitude or prior experience, allowing them to take on new responsibilities more quickly than others in operations.

⊕ Sharpshooter Tips

- Always champion opportunities to align your team toward core fundamentals of risk and business priorities. It doesn't have to be complicated or expensive. Make it fun, collaborative, and safe to participate for all while championing a culture of teamwork.
- Never underestimate the power of a united team, even as small as two. While some may argue it's inefficient, a team that trusts one another and works together without drama or division is far more effective than individuals who compare, compete, and backbite.

⌐͟͟͞ Misfires

- Authoritarian rulers breed resentment, isolate their staff, and increase anxiety and fear, resulting in division and a toxic culture.
- Shame-based cultures destroy growth, taking risks, and community development. "You should already know that" is not your friend! Instead, a culture of continual learning should be fostered based on mentoring between expert and apprentice. Make it safe so taking risks and attempting new skills and abilities are rewarded and encouraged even when things don't turn out and mistakes are made.

DESIGNING ROADMAPS WITH BALANCE

Designing roadmaps with balance and wisdom is essential for success! If it's something new to your organization, start small, learn the ropes, and gradually scale up once you've proven its effectiveness. For example, begin by implementing Zero Trust Architecture on a noncritical asset with a limited scope before attempting to scale it upwards. Roadmaps are an excellent tool for implementing follow-up after establishing a cybersecurity framework. After baselining operations, identifying gaps, receiving recommendations from third parties, and holding internal discussions on priorities and next steps, create a plan of action. Ensure immediate next steps are taken to achieve quick wins with low-hanging fruit. Then, prioritize items for the near term while planning and staging for those that must be delayed, ensuring they are positioned for future success.

₴ Everything is important!

In leading a global cybersecurity framework practice for F100 organizations, I assisted on an engagement where the CISO took operations from one person supporting the organization and, after multiple mergers and acquisitions (M&As), expanded to a large global SecOps team. As part of the engagement, I privately asked the CISO for his initial perceptions of where the organization was strong and where they were weak. I also asked about his expectations for each control category. I then promised to share the initial data with him and level-set his expectations and perceptions against the reality of the baseline evaluation. I then asked the CISO to prioritize a few key security areas critical to us as consultants so we could focus our engagement, research, and development efforts on reporting and the roadmap. To my surprise, the CISO said, "Everything is important!" instead of providing a prioritized list as requested.

Sharpshooter perspectives

SSP1: This may be a case where the CISO is incapable of truthfully ascertaining the current state of security and needs the consultants to provide a dose of reality. As a consultant, I've faced this as well. A large organization I met with had three CISOs, all responsible for different areas of the business, yet the common services were shared. Each had their perspective of security for the standard services, and their perspectives were very different. All three couldn't be correct, and after further investigation, it turned out that all three perspectives were incorrect. During the initial findings meeting, reality set in, and all three CISOs understood how their perspectives were incorrect and began working together to align their programs. As the consultant, you are the expert and must decide what is important and tell the client.

SSP2: The CISO is correct; everything is important, yet some things are more important than others. I've heard CEOs, CIOs, and internal audit leaders express the same sentiment. To educate them, I have often thrown in an obvious area that could be more important. For systems, I might have something like (1) a document management system (DMS), (2) an email system, and (3) an employee goals tracking system. Suddenly, a light goes on that they cannot live without email and that the DMS could be down for 24 hours, but tracking goals could wait for weeks! The critical point is convincing them to categorize one item, which usually leads them to make decisions in the other areas. A CISO who feels everything is important is a CISO with more priorities than they can take on, which will lead to burnout.

SSP3: I've heard this repeatedly, and it's a giant red flag indicating that leadership may be uncomfortable with the accountability and responsibility that comes with their role. If everything is a "priority," then nothing is.

Target grouping

As a consultant, I recognize that our engagements often involve a journey of trust, transformation, and facing hard truths as the engagement with the client unfolds. During the initial interview, it became evident that the CISO had taken a "small shop" mentality (hero complex) and tried to manage a massive enterprise with many stakeholders after several M&As. He was now overwhelmed and didn't know how many or what types of devices were spread out across the enterprise. He also had no control over the security components dispersed throughout the infrastructure. This occurred while he clung to his former sense of control and "hero-itis." He had been so successful in just a few years that he was blinded to the massive risk quietly accepted through M&A without proper oversight or risk management. He was essentially on an episode of "Naked and Afraid" without realizing it.

We worked to build trust and gently come alongside the leader, hoping that the data and the process would force his hand, and it did. He had been a great leader who had lost his way through M&A, which we believed would be his saving grace if we had the courage to champion him in this situation, which we did. Once the assessment data arrived, we revealed many gaps showing the substantial difference between his perception and reality. We let that settle in to address our larger objective: priorities. How could we get him to prioritize the objectives needed for this program?

The following week, we asked the CISO to prioritize the top threats. As mentioned in our opening story, he responded with, "Everything is important," essentially refusing to prioritize the list. I met with him on a private video call and asked him to entertain me by doing his best to rank the list numerically. He agreed, giving me hope. Several days later, he provided a prioritized list and admitted it was more complicated than he thought. At this point, I knew he understood his perceptions were off, he lacked priorities, he was overwhelmed and out over his skis, and he needed help. I spoke with him about the challenges of being a CISO, relating to him emotionally, and addressed how CISOs must forge through this with extreme clarity and how to achieve that.

We recommended a long-term roadmap and short-term actions, focusing on a framework-driven set of priorities. This started with the foundations of security solutions around visibility (CMDB and data classification/inventory) and proactively managing M&A risk, possibly slowing it down as it significantly increased their risk. They needed help, so we swooped in, created policies and procedures, and helped them operationalize to a level three CMMI process-driven state. This enabled them to achieve quick wins to impress the board while taking immediate action on the risk roadmap and building trust in a new direction with clear priorities.

⊕ Sharpshooter Tips

- Clear priorities are critical as a CISO. You're ineffective if you have more than 3–7 priorities. Don't try to boil the ocean; you will not be successful!
- In the CISO role, where everything can seem like a priority, identify your priorities by distinguishing between short-term and long-term goals. Focus on quick and easy wins and the most critical outcomes first, ensuring you don't create any roadblocks for subsequent goals and outcomes. Follow the priorities staged throughout the year according to your roadmap. You should delegate project management to provide regular oversight and management of objectives, outcomes, status, and changes. This ensures adaptations, targets, and adjustments are made for overall success.

⌐⁷ Misfire

- Don't be a burnt-out CISO who does too much and becomes ineffective and exhausted while accepting risk in all the wrong places.

己 I automated everything; Why do we need GRC?

I was working with a client on maturing their SecOps using a cybersecurity framework when the Director of IT said, "I automated everything to meet all these controls; why, then, do we still need governance, risk, and compliance?"

For context, we rated their maturity as a two, typical for the initial operations baseline maturity score based on the CMMI model, which has five levels. We recommended they achieve a level three industry-standard best practice where operations are defined, documented, and managed, and everyone knows where the documents are and follows them. The CMMI model has three levels considered Immature: (0) Nonexistent, (1) Ad Hoc, and (2) People-dependent. Upon reaching level three, the organization becomes process-dependent with well-defined operations.

The CISO had promised board members that they would achieve levels (4) Quantitatively Managed and (5) Optimized in some areas. The Director of IT said he had studied the CMMI model and understood that level four involves measured and controlled processes, while level five focuses on optimizing process improvement. He asserted that he had already achieved these levels within his technology stack by configuring detections, alerting, reporting, and logging.

Sharpshooter perspectives

SSP1: In the CMMI, the detection, alerting, reporting, and logging aspects are primarily associated with level 3. Organizations establish and maintain robust processes at this level that enable consistent and effective product and service delivery. The Director of IT needs to be corrected and educated since they have a long way to go to reach level 4, where processes are consistently measured and controlled. At this point in their maturity journey, they can only dream of reaching level (5) Optimizing, where everything is so well orchestrated that the focus can be process improvement.

SSP2: With level 4, it's more than just automation. It also uses data to manage the process. It's fantastic to have automated so much; it could be a step in the right direction. The next question would be how the data is used to manage the processes. I have seen vulnerability management analysts automate the process of sending results into ServiceNow for remediation without examining the root cause of the vulnerabilities. This approach helps manage and streamline the process but misses opportunities for additional improvement.

Target grouping

Perhaps the only internationally known and accepted word is "no" – all others, we struggle to achieve a common understanding, which was the case in this situation. From a technology perspective, the Director of IT read the CMMI documentation and assumed all the levels could be achieved through technology while not understanding governance, risk, and compliance (GRC) and how that was accomplished. He was out over his skis and didn't realize it. Worse, his ego, pride, and desire to control and build up his domain of influence as he grew out his team and budget made him obstinate and stubborn, unwilling to listen to those who knew differently. Several difficult and delicate discussions took place to clarify and articulate the differences between perception and reality and to explain how levels four and five maturities are to be achieved, but to no avail. The contract progressed from a good start to excellent, only to come to a screeching stop due to the ego and pride of one misinformed individual. This person will eventually learn the hard way from an auditor or cybercriminal.

CMMI is an international standard for understanding maturity and an outstanding resource that every CISO should embrace in SecOps. Most operations are typically around level 2, with some documentation but inconsistencies. There is usually still some tribal knowledge and dependence on people who are considered "heroes." Level 3 is the industry standard, where organizations move from ad hoc and people-dependent processes to process-dependent operations that are consistently followed. Achieving levels 4 and 5 is very challenging. These levels are more data – and risk-driven, relying on governance that predictively reduces risk with outstanding prioritization and efficiency. Due to the expense, levels 4 and 5 are typically desired only for the most critical operations and risk reduction areas.

⊕ **Sharpshooter Tip**

- Structure yourself for success using proven international solutions and standards like CMMI to drive outcomes and results.

◿ **Misfire**

- Being arrogant, stubborn, or ignorant and rejecting expert advice is foolishness at best and worthy of termination at worst. Every strong leader must be confident, bold, courageous, and willing to take risks while being careful not to become prideful.

⌇ These are the priorities!

While consulting at a large Fortune 100 financial organization preparing for a merger and acquisition, it was critical for them to achieve and demonstrate maturity in key areas, which would allow the team to keep their jobs during the merger. However, there were a lot of "cooks in the kitchen" on this project, creating several challenges:

- Complex politics
- High pressure in a merger and acquisition environment
- Various contractors in the environment that had personal relationships outside of work
- Challenging and toxic culture
- Changes in staff and priorities as the company prepared for the merger.

In managing the project toward specific goals and outcomes, we attempted to prioritize each, leading to many challenging conversations and nonconstructive feedback. There were discussions about individuals seen as roadblocks or not aligned, and some team members did not understand the big picture and overall company needs. Unity did not exist as the pressure mounted, making it critical to align priorities and the team toward common goals. Clearly defined and scheduled short-term and mid-term projects and projections were essential to ensure success within a tight timeline.

Sharpshooter perspectives

SSP1: This is where an authoritative leader needs to emerge. There's no time to get everyone's opinion and come to a consensus. I've worked in M&A, and it's more like an incident, so treat it as such. The leader needs to set the vision, communicate it, break the work into manageable segments, assign clear responsibility and authority for those segments, and hold those assigned accountable. There will also need to be some overarching goals that incorporate large groups of people, and the reward structure needs to be built for those areas so everyone must succeed, or the reward system will not be activated.

SSP2: I learned this one the hard way through a merger. First, determine who can deliver and who is all talk. You will have to fire the people who cannot deliver. It is nothing personal. The hardest thing you may be told to do is to cut staff to meet merger goals. Determine what the organization will lose in the cuts and communicate this. Unfortunately, you will also need to accept that everything will not be perfect, which means that if someone is delivering but has some flaws, you may have to live with the result and fix the issues later.

You must communicate a vision. Don't discount what the other company you are merging with has to offer; they may be better at some things than you are. Do not play favorites – base recognition on the delivery of objectives.

Watch for individuals who work behind your back to undermine the vision and determine if there is a basis for their criticism. Cultivate a culture where the critics can say things directly to you. The critics who genuinely care about the organization will say it to your face. Listen to them and determine if what they say has merit. However, in the end, you will still need to make the call.

If a critic goes behind your back, it means one of two things: (1) you are not approachable, or (2) they are jockeying for position. Figure out their motivation because, as a leader, you must be able to work with both types. Get used to the fact that people want to move up. Most importantly, fire individuals if they are not delivering.

Target grouping

The other day, I saw a meme and short video that rang true for this story: life is hard, and if you avoid hard choices, the consequences are more complex, so make hard choices today! What hard decisions need to be made here? Personalities and politics need to be removed immediately. Decisive leadership must step into a direct role and hold people accountable. During M&As, removing those who are underperformers or not a good fit is essential. Organizations need individuals who align with their cultural and leadership goals and do not spread toxicity.

In solving challenges in this environment, I leveraged what I call the "Prioritizer," a tool I have used for years in my practices and consultation. It's a simple Excel spreadsheet that documents people, variables, values, and ratings, as well as a few conditional color codes and numerical ratings to document and prioritize feedback from each person quickly. It's akin to whiteboarding coupled with variable tracking and risk metrics to qualify and quantify prioritization. Folks that don't feel heard get their thoughts documented in light, and they love it. Core leaders document what they see as prioritized. Collectively, we unite, analyze, and reveal how the team prioritizes.

As consultants, we provide our expert opinion to recommend or overrule decisions, particularly when the data suggest a different direction or when company bias or oversight causes them to miss the mark. This helps ensure they hit the target. Executives can constantly adjust based on what they see and must have the final say. This approach worked brilliantly, making every-thing process-dependent, ensuring all voices were heard, documented, and progressed based on data rather than individual opinions.

⊕ Sharpshooter Tip

- Unity is not easily achieved. Always be aware of what drives and moti-vates each person. Some are insecure, others are jealous, and others are scared. Identify what drives a person emotionally, relationally, politi-cally, or otherwise to determine the best way to navigate the complex political landscapes wisely.

Misfire

- Don't become part of the problem by picking sides or joining "he-said-she-said" conversations.

Chapter 4

Defending the enterprise

People often assume that a technology is secure or risk-free because it is widely used in the industry. As a CISO, you will constantly push against this attitude as the latest technology enters the company. We have evolved vulnerability management systems to track missing patches and configuration issues. In-house development presents a tough challenge due to an increasing range of new technologies and development platforms. In addition, most enterprises have multiple external vendors who supply everything from HVAC to network storage. The typical CISO must work in various areas, from understanding the regulations on specific types of information to negotiating security requirements within vendor contracts. CISOs in regulated environments are commonly audited by internal, external, and client audit teams, constantly fixing new findings as they arise. Defending the enterprise is a challenge as the company establishes new systems and new regulations are enforced. Most CISOs will ultimately face an incident that impacts their firm, testing their mettle and forcing them to use all of their talents to drive the attackers out. This chapter focuses on managing risk, operations, and incident response when all else fails.

MANAGING ENTERPRISE RISK

The CISO must understand the technology in the environment and its link to business strategy, enabling the strategy while balancing regulatory and security risks. This section examines several stories of governance, audits, and how to approach managing enterprise risk without bringing the business to a standstill.

⧖ Marketing form: Why can't we do this too?

I was a CISO at a financial services company, and one of my senior security engineers pulled me aside to have the following conversation:

- Engineer: "You need to tell the marketing group 'No.' They want to put a form on all of our external websites that allows anyone to enter an email address and use an open form field to email a 'Hey, I just saw this great investment,' with a link."

DOI: 10.1201/9781032720500-5

- Me: "If many other companies have this functionality on their websites, marketing will immediately ask us if someone else has a form like this; why can't we do it?"
- As the conversation proceeded, it became evident that the engineer had already encountered this resistance when he tried to tell the marketing group that they could not have the functionality they wanted. So, we had to take a different tact.

Sharpshooter perspectives (SSP)

SSP1: As a CISO, it is important to understand external environments that may impact your colleagues' perceptions. This understanding allows you to ascertain what they will ask for, how they may justify risky behavior, and be ready with an alternative solution that balances the business opportunity with security risk. Approaching situations in this manner helps security be part of the solution, not the "Department of No."

SSP2: This situation presents a great "Teachable moment" for the engineer. The engineer's approach is just to say no. By coaching them to consider alternatives that may work for marketing while being secure, the CISO can help expand the engineer's critical thinking skills. The CISO can also set the expectation that the engineer has developed at least one alternative for resolving the next challenge before bringing it to the CISO.

SSP3: A CISO isn't just a technology professional; they are a business leader. With that comes the responsibility of knowing the business objectives and strategies that enable the organization's growth and longevity. Sure, marketing is one of your primary sources of internal threat, but their role in the organization is essential, and there is often a secure and compliant way of getting the results they want. It is often better to take a "No for that option, but let's get to a yes that works for both of us" position and be seen as an enabler rather than a blocker.

SSP4: The leader is to champion their key enablers and requirements with the marketing and branding/PR leader to ensure successful operations. Alignment at the top is critical. If leaders allow culture to be determined from the bottom up, everything will be turned upside down.

Target grouping

Marketing had just asked us for a form that an external person could use to email any address. We know other companies use the same functionality, but how could we show marketing how to implement a form with reasonable security controls?

The engineer and I sat down and drew up a list of our competitors in the space, and then we went to all of the companies' websites. About 60% had the functionality to email a link to someone about a particular investment anonymously. The interesting thing about this approach is that we could send links to ourselves from other companies and examine the emails we got back.

We did not perform penetration testing on any forms and instead used standard email addresses to examine headers. The results gave us a lot of information on how other companies had designed their forms.

None of our competitors had a secure design. For example, some appeared to be susceptible to injection flaws. Others sent email from their main email domain, meaning that if someone were to use the functionality to spam someone else, the entire company's email domain could be blacklisted. We could see internal servers and IP addresses in the email headers we got back. Most of the forms also did not appear to rate limit requests. Using this information, the engineer and I constructed a set of design flaws and the initial recommended remediation, such as:

- Risk: Many of the forms appear to be susceptible to injection flaws (OWASP A03).
- Remediation: We can reduce this risk by eliminating the free text form and replacing it with a standard message, like "I thought this investment was interesting." Note that we will still need to test this for injection flaws.

All of this information was packaged into a presentation that we gave to marketing. Essentially, it was broken out into:

- Here is a screenshot of our competitor's form.
- Here are the design flaws and risks.
- Here is the recommendation so that we don't make the same mistake.

Overall, the presentation moved the conversation around the risks and how to control the risk rather than just giving a flat "No, you cannot do this." As we gave the presentation, I was growing more concerned. Although the marketing team listened intently to what we were saying, they were very quiet. Toward the end, the first question came out, "Have you looked at this particular competitor? Because we know the marketing team there."

I now knew that the marketing team was interested and had absorbed what we were presenting. We looked at that competitor, and they had the design flaws like everyone else. Ultimately, the team accepted our recommendations and considered our site more secure than the competition. In addition, the marketing team began pulling the security team into development conversations early to help manage future risks.

⊕ **Sharpshooter Tips**

- Most business leaders are extremely interested in the competition. As a CISO, the more times that you can provide examples of how the competition operates and potential mistakes that they are making, the more likely the other business leaders will pay attention to you. As a CISO, be prepared for the question about how the competition operates and note which other companies your business leaders feel are competitors.

- Adopting a "Yes, and here is how we can do this..." approach can be more valued and respected in enabling business strategy while managing risk.

🖊️ Misfire

- When a competitor has a specific functionality on their website, it is nearly impossible to tell your business leaders they can't have the same functionality to keep up with the competition!

References

- https://owasp.org/Top10/A03_2021-Injection/
- https://owasp.org/Top10/A04_2021-Insecure_Design/

⊇ An unpopular decision

I was the CISO for a company where personal webmail (email via browser – aka Gmail) was allowed. Executive management loved that they could check their personal email from their work computer throughout the day. However, I was very concerned about webmail being used as a conduit for insider data theft, but internal auditors never raised the issue due to its popularity. As the CISO, I was consistently in the crosshairs of our clients, who were understandably worried about their data being stolen by insiders using personal email. Unfortunately, raising the security risks found little traction in upper management.

As a security team, we decided to work around the vulnerability. We deployed network malware detection devices, endpoint data loss prevention, incident detection, and response tools. One network issue we had was that most webmail is encrypted. Although we could decrypt the email, we knew it was not a viable option for privacy reasons and would present regulatory issues for an international company. I pulled together incident metrics for the board and executive management each quarter, including compromised workstations. The most recent metrics showed that five of our last seven compromised workstations had been someone clicking on a phishing email from their personal webmail on the company workstation. The executive briefing would be memorable as I showed them the metrics.

Sharpshooter perspectives

SSP1: You've got the data needed to push for blocking personal email. Before the meeting, I'd go to any established allies on the board and in executive management, provide a personal brief tailored to their biases, and get their buy-in for the policy change. My storyline would be something like this:

- Our largest clients, which generate $x or x% of the company's revenue, are becoming increasingly concerned about their data being stolen from us through our employees, who would use personal email to exfiltrate the data.

- The threat of compromising our environment, losing the ability to operate, and simultaneously losing the goodwill of our clients is real, and the number of compromises from email during this last reporting period shows the probability of compromise has increased significantly.
- By eliminating personal email from our environment, we can meet our most prominent clients' expectations and significantly reduce the probability of a breach. Recognizing the importance of personal email access, we will launch a campaign to help staff access their email on their smartphones at all times.

SSP2: Block personal email and personal file sharing. It's common sense these days and, honestly, a no-brainer. Sure, having that available is more "convenient," but that is what your personal phone is for, while the business network and workstation are for business use. People may not like the answer because it is not convenient for them, but then again, having the entire company go out of business or get ransomware because you wanted to do personal things on corporate assets is also not easy to explain to regulators, shareholders, and customers. Believing your ease is worth damaging others' trust, privacy, and possible livelihood is the height of entitlement and selfishness.

Target grouping

I was part of an industry group that sent an annual security questionnaire summarizing the controls across the industry. As one of the committee members who created the survey questions, I had the opportunity to add the question regarding who had webmail still open. The survey results showed only 10% of companies our size had open webmail. In addition, one of our larger clients recently asked us to close webmail, deeming it an unacceptable risk. Finally, I had the data regarding the malware incidents regarding personal email.

I presented my recommendations to executive management during the metrics briefing. Despite some heated discussion, the CEO ultimately decided that my team should block personal webmail.

⊕ Sharpshooter Tips

- This story combines several tips for advancing security controls when executive management is only partially on board. A CISO should maintain valuable statistics indicating the root cause of the incidents. The incidents in this story showed personal email as the root cause.
- A CISO should maintain good information on clients' issues regarding the company's security controls. In this case, clients had expressed issues with personal email being open.
- A CISO should maintain ties with industry groups. In this case, the industry group provided excellent survey information on how the company's security controls compared to its peers. This information

was highly relevant to the current risk because it directly related to competitors.

- Modern browser isolation technologies can be used in this situation to enable individuals to use their personal email while maintaining a separation from the corporate network.

⮂ But It's a legitimate company!

Most firms have software that blocks specific sites based on categorization, such as phishing or malware. Inevitably, as a CISO, you will eventually have a conversation like this:

BUSINESS ANALYST: "I need you to unblock this website. I need it to perform research on this particular company."

CISO: "The website is classified as having a drive-by-download; if I let you go to that site, you may infect the network."

BUSINESS ANALYST: "But it's a legitimate company! I have been talking to them, and I need to go to their site to get their information."

Within my Information Security Team, we created a set of isolated environments to analyze malware sites. The environments were beneficial for when someone had clicked through a phishing attack. I informed the analyst that we would examine the site and determine if the classification was wrong. Less than 30 minutes later, one of my information security analysts pulled me aside. Embedded on one of the inner pages of the legitimate company's website was a defaced page with an image of the character Ryuk from the anime Death Note and an inscription that read "hacked by" with the hacker's name. Hackers rarely leave a calling card because it gives away the fact that the site is compromised. I asked for a screenshot and scheduled a time with the business analyst.

Sharpshooter perspectives

SSP1: Your security team did a great job. You are doing exactly what you should be doing: protecting the environment for everyone. A single business need does not override the security of the entire business. You proved through a best practice – a secure, isolated environment – that this was a compromised website. Even though this is a legitimate client website, it should remain unavailable until rebuilt and secured. The business analyst should contact the client and let them know that any work associated with the website needs to be delayed since the website has been confirmed as compromised, and nothing on the site can be considered legitimate.

SSP2: This is what's supposed to happen. Test don't trust. The security work here showcases the right way to "stand your ground" and take the right shots to either prove or disprove the legitimacy of an assertion (in this case, that a "legitimate" site couldn't possibly be malicious). Nice work.

Target grouping

A business analyst found that the company's legitimate website they were researching was blocked. The security team's investigation found that the website had been compromised and had drive-by-download code that would attack systems connecting to it.

In the meeting, the business analyst reiterated that they needed the website opened up because it was a legitimate company. I asked if he had contact information for the company because I had definitive proof that the website had been compromised. I showed him the images and the URLs of the hacked pages. He was surprised but worked with us as we obtained screenshots through one of the isolated environments. In the end, he reached out to the company's contact to let them know that their site had been hacked.

⊕ Sharpshooter Tip

- Extreme caution should be taken even if the website is for a legitimate company. A CISO should have tools for its team to investigate malware, phishing, and compromised websites safely. These tools should be isolated from the central business network so that an investigation does not compromise the company. In general, the Information Security Team should have:
  - A malware sandbox to submit samples and websites,
  - An isolated browser to examine suspicious links and websites, and
  - An isolated workstation to handle samples.

⌐⌐⌐ Misfire

- Many companies do not have good security awareness programs, and typically, this results in the employee disconnecting their workstation from the network (if possible) to go to the forbidden site or using a personal computer to access it. In this case, the company had a good awareness program regarding malware, so the analyst's request was routed to the security team to address.

⊇ Oh, they left 2 weeks ago

My firm was going through a merger with another firm located on another continent, allowing the newly merged entity to dominate across both regions by solidifying operations. The firm stock would still be traded on an American exchange, meaning we must continue complying with Sarbanes-Oxley (SOX). The firm we were merging with was non-US, but they assured me they had stringent controls to meet US regulations.

One afternoon, the Deputy Chief Financial Officer pulled me aside and said, "Our external auditors have decided that we need to comply with SOX by the end of this year."

"Hang on," I said. "Didn't the auditors say we had until the end of next year to comply because of the merger?"

"They changed their minds," he said.

With a merger, one of my more significant concerns was the people streaming out of the door with layoffs. A central SOX control was that access was removed within one business day after a termination notice was received. I asked one of my analysts to take all the termination notices and link them to the deactivation date. As he returned, it was clear that several notices had been backdated, meaning the person had left days before the notice was sent. I called some of our remote offices to confirm with the IT units if each notice had an accurate departure date. A backdated notice had just come out for an office in Australia, and the call was chilling: "Oh yeah, she left the company 2 weeks ago. She didn't know what to do with the equipment, so she dropped it all off yesterday."

Sharpshooter perspectives

SSP1: An effective and efficient termination process is essential to protecting the company. The process typically requires a close partnership with HR or whoever is tasked with terminations. In some places I've worked, the project manager oversaw terminations. I'm concerned with any backdating, which is an inaccurate and falsified record at worst. Since the merger had recently happened, it would be safe to say there is no centralized identity management to turn off all access upon termination quickly, but that should be a goal of the organization. If the decision is to remain decentralized, then a centralized termination policy is essential, with implementation procedures required for each area. The process should be coupled with a campaign to educate those responsible for terminations to understand the risk and the importance of removing access immediately and having equipment returned expeditiously. Like many things in maturing a cybersecurity program, this one will take time to get under control fully; however, having documented action plans, milestones, and status reports can go a long way with the auditors.

SSP2: The importance of a strong IAM program cannot be stated enough. ANY organization with ANY data that could be considered confidential in ANY fashion needs to have this function on lockdown. The "Joiner," "Mover," and "Leaver" processes and automation are foundational to understanding the proper assignment and removal of rights when people join, move within, or leave the company. A frequently missed one is the "Mover." Proper authorization changes are not just a function of joining or leaving an organization, but also, when an individual changes roles within the company, you must ensure that the new access rights are limited to the new role. They don't get any "carry-over" rights from the last role – especially if they had more access. The implementation of "least privilege" is critical here. Another item of note is the close integration with HR, which needs to verify this occurs.

Access changes must be integrated into the ecosystem, not a collection of isolated processes.

Target grouping

During the merger, the HR department began to backdate termination notices, often sending them days after the employee had departed. The improper backdating violated the termination policy and caused the SOX controls to fail.

I brought the information to the HR executive, and he was unphased. He explained that mistakes like this would occur. I explained to him that these errors would likely be considered a "material weakness" to the auditors in our termination processes and be seen as a "significant deficiency in the internal controls over financial reporting." The executive was indifferent to the issue since the control was structured to place the blame on my team that managed access, including removing terminated employees. Despite the termination date being backdated, auditors would leverage the backdated record to demonstrate that my team had failed to deactivate the account within one business day of the termination notice.

I escalated these findings to my management and HR. I needed to know if the same external firm was conducting a different audit before the SOX audit. As I addressed the backdated problem, the external firm discovered the backdated termination notice from Australia. This finding and several others led to the threat of a "material weakness" in the controls. Since the period they were testing for the other audit overlapped with SOX testing, the implications were severe: the company could also have a failing control for SOX compliance.

The finding culminated in a meeting between me, the HR executive responsible for terminations, the CIO, and the VP of HR. The first thing the VP of HR said was, "This is the first I have heard of this." The HR executive looked very nervous, but the VP of HR didn't reprimand him, which meant that his behavior and the process were unlikely to ever improve. I did hear that the exec was recently laid off, so maybe it did catch up with them.

Ultimately, I was provided with resources and funding to hire an external consulting team to improve access management. Because the termination controls were failing, the external auditors emphasized the access documents and informed us that the termination controls would be considered a failure for the year.

For the next 6 months, I worked with the consulting team to collect reports on the system access logs and have them reviewed. In addition, I received a resource to help install a new system for additional access reviews. The merger complicated the process, as the system owners had to familiarize themselves with all the personnel from both companies. However, one positive outcome of the remediation effort was the enhanced focus on access control and its improvement for future security.

⊕ Sharpshooter Tips

- Auditors often blame the access management team for departure failures. A CISO should work closely with HR to define the termination process. There are a couple of ways to improve this process.
  - Automate the process to remove terminated employee's network access.
  - Separate the regulatory control into two pieces: (a) HR notifies IT before the termination date to deactivate the account at a specific date and time; (b) IT deactivates the account within one business day of the termination.
  This separation process clearly defines accountability, ensuring each team is responsible for its respective parts of the process.
- Mergers typically generate an increased volume of departures. The CISO should ask for additional resources to handle the increased volume and encourage HR to do the same.

⌐⇒ Misfire

- Mergers that consolidate and lay off employees often lead to contentious departures. Additional HR and IT should be brought in to manage the increased workload. Additionally, a specific plan regarding regulatory compliance should be established to ensure all stakeholders are aware of the timelines.

🐍 Throw all your logs into the SIEM!

One of the firms I worked for had an extremely thorough audit department. The auditors regularly hired external firms to perform penetration testing within the company and brought in experts to conduct audits. On the one hand, external experts often knew what technical questions to ask to audit the security of a system. On the other hand, the external auditors behaved like hired guns, which kept the security team on edge. In a recent firewall audit, the auditors provided my SOC team with a spreadsheet of 5,000 IP addresses, stating, "We pulled the past week of traffic of people surfing the web. These IPs appear to be for unclassified websites, and you need to tell us that these sites are not malicious." URL filtering firewalls generally have a list in which they tag the URL with a classification; for example, Fox News would be classified as a "news site." Unclassified sites indicate that they are either new sites or not heavily utilized. It also could mean that the site had been recently created to host malware.

If we could not provide the auditors with the information, they would write a finding that our SOC team should be reviewing all unclassified traffic or that the company should restrict all unclassified sites. The challenge was that the firm had many analysts who researched new companies with new websites. Locking down all unclassified sites would have a significant business impact.

Because of the volume of work, I joined the SOC team to review firewall logs and pull the URLs for each IP. The work was tedious and took several days of research. The audit team was impatient and scheduled a meeting to have me present what our research had revealed.

Clearing my throat, I started on the first IP address, "This is an employee who went to a small business site; the website was Comfy Nails." Then I went to the next one, "When you look at this traffic, this is an employee pulling the menu for a restaurant called The Kitchen." After about the tenth IP address, the team had me stop rattling off URLs for local restaurants, bars, and spas. I thought that that would be the last of the firewall audit until a week later when I got the audit report. The auditors discovered that the security team was not sending any external internet-facing firewall logs to the SIEM for correlation, which they felt was a gap in our controls.

Sharpshooter perspectives

SSP1: I guess the audit team would identify a finding no matter what, so they made one up. The key term "not sending any" regarding external internet-facing firewall logs going to the SIEM is very telling. I expect that critical event logs will be sent to the SIEM. I'd push back like when my client, the Department of Energy (DOE), had an unannounced, 2-week external pen test. At the close-out meeting with the senior leaders from the DOE and my company, the head auditor stated, "We were able to breach the external firewall and gain access to the internal network." I was in shock, knowing that if this were true, my sensors missed it, and there would be a lot of reputational harm for me and the team, which was about to occur in the DOE community. And why didn't the lead pen tester, whom I had built rapport with, let me know so that I could fix the issue immediately?

My gut feeling indicated this was bogus, so I stood up and calmly addressed the room. "All our technical indicators say this finding is inaccurate. Let's call the lead pen tester and ask." There was dead silence. How could they say "no" since I was already calling using the conference phone? Luckily, he answered. I asked him, "Were you able to breach the external firewall and gain access to the internal network?" He said, "Absolutely not. We were detected and shut down, so we had to identify ourselves and divulge the audit. We could still not gain access after formally acknowledging that we were stopped and allowed to continue unabated." The audit team was reporting to Congress the next week and had already created their presentation and assigned our site an "F" even before the audit occurred. They had to admit this report was inaccurate. Two years later, the lead auditor sought me out at a conference and told me that due to that audit and the pressure he had been put under to provide predetermined results, he quit that organization 2 months later for integrity reasons.

SSP2: An auditor's job is to find something, and they always will. The best way to approach this is to establish rapport with the auditors and help them find what you want them to find. Unless the auditing team are complete ass-holes (technical term), they can be fantastic allies to focus organizations on the areas you need additional resources and attention. Listen, you know your program isn't perfect and never will be. You also see where you could use more resources to help reduce organizational risk. Help the auditors do their job in finding issues and have them help you showcase the things you need help with resourcing to resolve. It can be symbiosis.

Target grouping

During an audit where the auditors recommended that all external firewall logs be sent to the SIEM it became more clear that the audit team was intent on issuing a finding, even if the vulnerability was of negligible risk.

Unlike other companies I have worked for, this company had decent seg-mentation and two firewalls. The external firewalls performed standard IP filtering, screening millions of events daily. The second layer of firewalls was a different vendor, was application-aware, and also functioned as an intrusion detection system (IDS). Traffic from the second layer of firewalls was sent to the SIEM, while none of the traffic from the external firewalls was sent to the SIEM. Typically, if an attacker triggered alerts on the second set of firewalls, the SOC team knew that an attack had gotten through the external firewalls.

Anyone managing an external company firewall knows external probes constantly attack it. Knowing there was a second firewall layer in the defenses, I worked on several calculations if we sent all the external logs to the SIEM. First, I pulled the number of incidents in which the SOC team had used logs from the external firewalls. Because the external firewalls mainly dropped traffic, the number of incidents directly using those logs turned out to be 1 in 998 incident investigations over a year. The investigation pulled the local logs on the external firewall for the single event that utilized the logs. I calculated the amount of data and determined that sending all the logs would increase the SIEM storage costs by hundreds of thousands of dollars.

The audit team found the analysis reasonable because external logs were rarely utilized. They accepted my explanation that they were low-fidelity and that adding them to the SIEM would not provide any value.

⊕ Sharpshooter Tip

- Working with an audit team can be challenging sometimes. If they iden-tify an unreasonable finding, pulling together statistics based on actual threat and incident investigations can help make a case for proceeding with your response to the finding.

⌐ Misfire

- In this case, the internal audit team had a history of hostile interactions with IT and Security, which often caused friction between the teams.

Maintaining a good relationship with audit teams is critical to advancing a security program.

OPERATIONS

Most CISOs handle operational tasks related to governance, such as overseeing penetration testing and security awareness training. They also manage incident response teams that monitor and analyze alerts from various devices across the network. This section delves into vulnerability scanning, penetration testing, and managing an external vendor to perform incident triage. The stories provide methods to increase your team's reputation for excellence and position them as a trusted advisor to the business. The section ends with security awareness, which is critical to all programs.

己 Duck, here comes another grenade!

When taking on one of my first CISO roles, I discovered that Operations was patching some issues, but there wasn't a comprehensive program. Patching seemed to vary based on the Server Administrator instead of a process. I purchased a low-end vulnerability tool, Nessus, and set it up on a dedicated laptop to train a young Security Analyst how to use it. I explained the following concepts that every vulnerability scanning program should follow:

- Obtain proper authorization before scanning. For the Security Analyst, that meant written approval from the CISO.
- Avoid denial-of-service (DoS) conditions by limiting the IP range being scanned, understanding what should be on the IP range, and avoiding heavy scans over an extensive IP range.
- Share the results with Operations.
- Document the results and track them over time.

Given the demands of the new role and the immaturity of the program, I quickly focused on other priorities. I didn't think about it until I attended the CIO's staff meeting and learned that Operations was unhappy with security's vulnerability scanning. I left the meeting and headed to my office to call the Systems Manager. While walking through the hall, I overheard one of the Server Administrators say, "Duck, here comes another grenade." Not thinking much of it, I sat down at my desk and started reviewing the email and saw that the Security Analyst had just sent an email to that Server Administrator with another massive scan result attached with the comment, "You need to take this seriously and work on these vulnerabilities." It had been 2 months since Nessus was implemented, and the number of vulnerabilities had not decreased. Where did I go wrong, and what should be proposed when meeting with the System Manager to reduce the number of vulnerabilities and improve the relationship between Security and Operations?

Sharpshooter perspectives

SSP1: I wish this were uncommon, but unfortunately, it isn't. Security scans and assessments should never be used as clubs to smack the Operations Team and highlight their poor performance. That is a recipe for division and opposition. At its core, the vulnerability management (VM) function ensures that systems in operation are as resilient and stable as possible. It is a support function that assists operations in performing their jobs, which includes uptime and availability. If a system gets compromised because of a vulnerability, it is both a security AND an operations issue. With that in mind, one of the best ways to enable this is to get both teams in the same room and create shared metrics that everyone can agree to. There will always be vulnerabilities and finite resources to address them. When teams have joint goals and metrics, prioritizing remediation becomes more manageable. Additionally, understanding each team's constraints and focus fosters collaboration and strengthens bonds rather than causing division.

SSP2: I had to let one of my VM team members go because of a similar issue. I spent time coaching the individual and explaining they had to speak with the IT teams to get their feedback on the scanning program, but the Security Engineer wouldn't listen. He set up a script to automatically send the results from the scanning engine directly to the ticketing system but didn't prioritize, explain, or consolidate the vulnerabilities. For example, Chrome was missing 94 patches, so 94 tickets were sent to IT rather than one ticket to patch the browser. The IT teams ignored the tickets, and after a terrible audit, patching was identified as a corrective action. I asked a Process Engineer to help improve our processes, but the Security Engineer belligerently stated he didn't see any issues. After 18 months of asking them to communicate with the IT teams, with no improvement, I decided to let them go.

Target grouping

This experience taught me that a strong partnership between Operations and Security is essential and that significant progress is achieved through many small successes. I am also cautious of potential issues that may be seen as a grenade, taking extra time to explain the situation to all parties in delivery and fostering a perspective that we're all in this together.

The Systems Manager walked into the meeting with the mandate that the vulnerability scanning program had to stop immediately since the reports were so massive his team couldn't do anything about them. I agreed that the reports were daunting and the program was not working, so I proposed we temporarily stop the program and all focus on maturing the patching program. If we could make patching a routine operation, the scanning results should identify a more manageable list of vulnerabilities.

The Systems Manager expressed concern that patching could impact operations. I explained that the security risk of not patching had grown to a level that potentially rivals the risk to operations. After a series of discussions, we reached the following solution.

We put the vulnerability scanning program on hold to design a patching policy that we could agree to, including jointly implementing scanning as a patching verification tool when the infrastructure matured. To address the operational concerns, we patched test systems first, low-impact systems second, and high-impact systems last. Some systems were so obsolete that patching wasn't a valid option. We created an "Eliminate Obsolescence for Operational Resiliency Plan" for these systems and submitted it to the CIO, who approved it. The implementation progressed daily and took several years to complete.

Without input from management, the Server Administrator, who was already patching his systems, contacted the Security Analyst and asked that a few of his systems be scanned. They coordinated what systems would be scanned and when, and the Server Administrator took the reports and fixed the vulnerabilities. This Server Administrator became the champion of the scanning program and convinced his team that limited targeted scans were easy to manage.

⊕ Sharpshooter Tip

- Involve Operations in creating and implementing a vulnerability scanning and patching program. Approach it as a partnership, expecting a slow start that builds momentum with small wins.

⌐ Misfire

- The original implementation of this vulnerability scanning program was a misfire. Scanning was occurring, but vulnerabilities were not being fixed. A partnership with Operations is required to have a successful patching program.

⊇ Wow, your tools are really advanced!

Due to alert fatigue, I collaborated with my incident response team to explore using a Managed Security Service Provider (MSSP) for first-level support. The goal was for the MSSP to triage alerts and respond according to our established playbooks. Our team already had a SIEM tool, which fed information into an in-house analytics tool.

The analytics tool provided dashboards that allowed responders to triage events rapidly. The team had built custom automation into the tool to automate parts of the response to common attacks such as phishing and drive-by-downloads. Even with the automation, alert fatigue was a real problem, and we created a request for proposal (RFP) to find a vendor who could help us triage events. Most vendors inquired if they could feed the logs into their own

SIEM tools, doing away with the in-house SIEM. They interpreted the RFP as a request to provide the company with a SIEM tool rather than a triage service.

One local MSSP said they would work with the in-house-developed playbooks and feed alerts from our in-house SIEM into their incident tracking tool. The team signed an agreement with the MSSP to provide initial triage services. In addition, the in-house team provided the MSSP training on the playbooks and tools. On the first day, one of the MSSP analysts said, "Wow, your tools are really advanced!"

Sharpshooter perspectives

SSP1: It's hard to tell what is going to happen here. If all that is needed is staff augmentation and the MSSP is willing to adjust accordingly, this could be a positive solution. Hopefully, the MSSP will also bring additional expertise to strengthen your overall program. It sounds like the MSSP could benefit from the tools your team has developed, and it would be interesting to learn about the contractual language protecting your company's intellectual property (IP). Unfortunately, in my experience with MSSPs, their support models tend to be one-size-fits-all. As MSSPs grow, they often become more rigid, which can be detrimental to their customers. Even with a well-crafted written agreement, having a great experience with an MSSP is challenging.

SSP2: Identifying the right MSSP is critical. You have to be very clear about the expectations from the relationship and the specific tooling required to execute those expectations from the very beginning. I've personally been through this process a couple of times, and it often reduces to only a few viable partners to choose from when it gets right down to it. Many times, it's my tools, my data, my systems: your expertise, your effort, and your improvements in my environment. At the end of the day, the relationship can be developed to be great. Any maturity improvements can be made in your environment (I like to refer to them as evergreen artifacts). That way, when the relationship ends, for whatever reason, your program will be more mature and advanced and will not be left in a lurch with immature tech and processes that were never improved.

Target grouping

After hiring the MSSP to help the SOC team triage events and reduce alert fatigue, we found that we had more advanced tools and processes than they did. The MSSP actually reduced value!

We had signed a six-figure contract to reduce alert fatigue, and the next few months served as a trial period during which the MSSP demonstrated its operations. Unfortunately, they frequently sent us raw antivirus alerts, expecting us to investigate them. Any seasoned information security professional knows antivirus alerts can be routed directly to the SOC team without MSSP intervention. This pattern persisted across most alerts: the MSSP would log an alert and then instruct my SOC team to investigate without reviewing other available logs.

I met with the MSSP and explained that we needed their analysts to triage the alerts, such as analyzing the network traffic and processes, before sending us an event. My team successfully trained one of their analysts, but they were promoted and left the contract. After working with them for 8 months, my SOC team had lost confidence in their abilities because my team ended up triaging all of the alerts anyway, so I pulled the plug and fired them.

⊕ Sharpshooter Tip

- Firing the MSSP shocked several company leaders because I was admitting I had made a mistake. However, they were useless, so it was better to swallow my pride and remove them.

⌐⃗ Misfire

- Additional due diligence and customer interviews would have revealed that the MSSP was incompetent. After I let the MSSP go, another CISO told me they had also tried that same MSSP and experienced the same results!

己 He just asked for access?

One of our senior business leaders contacted my team to review a new IT system that one of our service providers wanted us to use. The service provider was a print vendor. Since we provided financial files to our clients, we needed the vendor to print and mail paper statements. Because of the sensitivity of the information, we had initially established a direct line with firewalls on both ends that were used to transfer files directly to the print vendor.

The vendor approached our business with a new IT system and asked us to move away from a direct line to an internet-facing file transfer system that they had developed in-house. Due to the sensitivity of the information, our business contacted the Information Security Team to evaluate the new system. The vendor provided documentation that stated, "This system utilizes state-of-the-art security technology to secure your files." On a call, I asked if we could get logon credentials to the system to perform our due diligence. We would need access anyway to upload files if we used the system.

I asked one of my Security Engineers to perform the due diligence. Since we hadn't received the logon information, he searched their website and found a portal to request a logon. He entered his name and company email. An hour later, he got an email stating his access had been provisioned.

A few minutes later, he said, "Come take a look; I think something is wrong." He was looking at a directory with thousands of files, which was odd because he had not uploaded anything. The first file was a company everyone would be familiar with, and it wasn't ours. Curious, he opened the file. The file was the company's confidential earnings report that wouldn't be released to the public for another 2 weeks. I said, "We need to step away from the keyboard and call the printing vendor."

Sharpshooter perspectives

SSP1: The incident needs to be investigated by the printing company to determine who had access to the different accounts and then inform their clients. Can the company whose files were visible get nondisclosure agreements signed by everyone who had access to those files? Will this prevent them from having to file a report with the Securities and Exchange Commission (SEC)? Probably not, but getting the signatures quickly and filing the report promptly shows the company is trying to maintain transparency. To protect your organization, search the files displayed for previous documents provided, and if any are found, open your incident investigation. I would discontinue business with this vendor and quickly establish business with another one you've vetted and has better security controls.

SSP2: Awesome due diligence, hard pass on the vendor. It immediately raises a red flag whenever I see something ambiguous referring to security (state-of-the-art security, proprietary encryption, etc.). If you can't specifically state something in objective terms that I can validate, it doesn't count. Security, privacy, and compliance all have discrete requirements and functional patterns. If those aren't there, my organization's data assets won't be either.

Target grouping

In performing due diligence on a system one of our vendors had asked us to use, we found that we had access to another company's confidential information. After we reported the finding to the external print vendor, they immediately threatened legal action against my company and asked my team to sign a nondisclosure agreement. Fortunately, while our legal team assessed the potential legal ramifications, the print vendor called us back, asked us to review the finding, and retracted their legal threat.

It took several weeks to investigate the vulnerability. We discovered that the portal through which the analyst had requested access was also used by third-party vendors in India to request access to the file transfer system. An individual was supposed to check and validate these requests, but they failed to double-check and mistakenly granted us access.

Disturbingly, we were given access as a standard support analyst. The print vendor explained that all support analysts were granted access to all documents in the repository at the request of their lines of business. We reminded the vendor that this violated the principle of least privilege. Their response was dismissive, essentially stating that this was how they would run their business.

Despite these issues, our business unit, which was still using the vendor, had no viable alternatives and requested to continue using the vendor. After acknowledging and understanding the risks, the business unit decided to continue with the vendor but did not switch to the new file-sharing platform.

⊕ Sharpshooter Tips

- Vendor due diligence is critical for managing business risk. This story illustrates that. Even though it was tempting to continue probing the vendor's system to see what else was visible, the ethical choice was to step away and contact them about the vulnerability.
- A solid relationship with other business executives is crucial for vendor due diligence. In this case, the business leaders were aware of the sensitivity of the information and skeptical of switching to a system claiming to have state-of-the-art security controls.

⌐⇁ Misfire

- If the security of the current vendor is in question, you should keep a list of alternative vendors. Unfortunately, in this case, the one business unit had no alternatives.

⊇ Do I have to do security awareness training?

"Damn right you do! Everyone was given 3 months to complete the 50-minute security awareness training, and you've been given four extensions. You've spent more time trying to get out of doing this than if you got off your ass and done it, not to mention how much of my time you've wasted. Also, there have been two incidents in the past year where you've clicked on malicious links, one of which you entered in your credentials. As a result, we had to change them. You complained that not only did you have to change your credentials for the company, but you also had to change them for all your accounts because it is too hard for you to have unique passwords even though you've been provided a password safe to manage them efficiently. Since you refuse to be minimally compliant, I will force you to complete the training the next time you log in and before you're given full access to your work environment. What? You've got a deadline that will take 48 hours to complete. Will it cost the company millions of dollars if it is not done and it's due tomorrow, so you can't have an interruption to access? As I made my daily rounds walking the floors, I noticed you've spent the last hour or so flirting with the secretary in the breakroom, so you must have some free time. Yes, I understand that you will use me and security awareness as an excuse to the CEO why you didn't get your project done. Good luck with that since I'll bring all the documentation of your incidents, all the notifications you've been provided, and the following reasons why security awareness is essential for the protection of the company. If you'll excuse me, I'm going to my office now to schedule a meeting with the CEO later today. I understand her 2:00 meeting was canceled, so an open slot should be available. I will also reach out to my CISO consortium to get some ideas on how to have a successful security awareness program."

Sharpshooter perspectives

SSP1: In every company I've been at, one to two people will do anything to avoid the training. The best way to get everyone to finish the training on time is to get approval from senior management to remove their access to anything on the company network except the training if they are x number of days late. Having an access removal process demonstrates the company's information security maturity. A less mature organization does not take awareness training seriously, and there are no consequences; therefore, people ignore the training. Many organizations are reluctant to pull someone's access, but regulators or auditors usually shift that perspective.

SSP2: One of the companies I worked for in the 2020s only had security training during new employee orientation, but none after that. Some people in the company had not taken security training in decades! In speaking with management, I discovered they did not want annual training because it would reduce the billable hours. However, the company was also under tight regulations, and the external auditors from our clients expected yearly training. After emphasizing that our clients required us to do annual training, I finally convinced senior management to require every employee to complete 30 minutes of security training.

SSP3: Everyone must complete annual compliance training for at least 30 minutes per year. If someone fails a control, like phishing, they must complete additional, specific training. But that isn't a proper training and awareness program, is it? A real training and awareness program should be designed to change your organization's actions and culture, similar to a social engineering influence campaign. It should be clean and focused, with design principles that make security, compliance, and privacy self-evident. These principles must be woven into the organization's DNA, so considering doing something without security becomes a radical concept. Training and awareness efforts are the psychological operations of your program and should permeate every engagement, conversation, and communication from your department. Training must be continuous for effectiveness, with ongoing repetition and focus, both overt and subtle.

Target grouping

All the quotes in the story above are things a CISO would love to say to those who haven't finished their security awareness training. We don't advocate this approach, but it was fun thinking about it. Regarding security awareness, let's look at what works and what doesn't.

What Works:

1. Obtain support from senior management on the security awareness program.
2. Make the program relative to employees' jobs.
3. Provide realistic timeframes for completing the training. Consider people with work assignment deadlines, vacations, holidays, and unexpected emergencies.
4. Communicate expectations with a consistent message.
5. Use carrots instead of punishment to encourage participation. At one company I worked for, it became a game to see who could complete it first. Everyone would print out their completion certificate and put it on the wall outside my office. When the company's top executive saw them, he printed his certificate and pinned it to the wall. Consider a raffle drawing for those who complete it by a specific time and have multiple drawings over time so that those who complete it first have more chances to win.
6. For those who refuse to participate, use a system that provides email and desktop notifications and increases the frequency of notifications over time.
7. For the most stubborn, route their login directly to the learning management system, so they must complete the training before gaining full access.
8. Allow a "test-out" at the beginning of the training so employees only view material they do not know.
9. Allow feedback at the end of the training and use that feedback as lessons learned for the next training. Acknowledge the students' comments. Sometimes, acknowledging that you've read and considered their comments will positively affect employees' opinions of the program.
10. Develop an engaging program appropriate for the audience. Some companies I've worked with found cartoon-based training effective, while others wanted something more sophisticated.
11. Make it company-specific whenever possible.

What Doesn't Work:

1. Publicly ridiculing someone for not completing the training or having trouble grasping certain aspects.
2. Long and in-depth training.
3. Generic training topics.
4. One training course for all positions.
5. All training is in person.

⊕ Sharpshooter Tip

- The weakest link in any cybersecurity program is the user, so have and maintain an effective awareness program.

☞ Misfire

- Making awareness training too long is always a misfire. It is better to keep the training to less than 1 hour and break it into shorter segments over time.

⅔ Personalizing awareness training

Security Awareness training is vital to an organization. In October, I worked with my Information Security Team to conduct a live training session for the firm. Individuals of all levels attended. In the first session, we discussed the technology in place to protect the company.

One of the Security Architects had established an external network, pulled the code for the Blackhole rootkit, and filmed a demonstration of the bot infecting a Windows machine. As part of the demonstration, he presented the bot's data collection abilities and how it transmitted information back to the command and control server. We also spent part of the demonstration discussing tools the security team used to detect and respond to security alerts like the rootkit attack. After the demonstration, we asked if there were any questions, and one of the employees asked an unexpected question. "You have completely scared me to death. Can you tell me how a normal person can defend themselves from the threat we just saw?"

Sharpshooter perspectives

SSP1: My answer would be, "Your best defense is to listen to and learn all you can from your security team. Follow our advice, such as maintaining your operating services up to date, patching your applications, using strong passwords and MFA, and having air-gapped backups if all else fails. I will drop you an email with some additional information where you can learn more."

SSP2: This training effectively used FUD (fear, uncertainty, and doubt). However, scaring people isn't enough, and the question posed is directionally correct regarding where a CISO should go. You need to meet people where they are and provide them with tools that match their level of understanding. The WIFM (What's in it for me?) needs to clearly show how this impacts them and what they can do to make a difference.

Following the security guidelines and processes outlined by the organizational security team is a start, but training on the why and how is even better. Provide them with a contact for support and the tools they need to succeed. In the context of security, everyone in the organization reports to the CISO, so treat them as the valuable resources they can be.

Target grouping

The Information Security Team had just provided a fantastic display of the company controls and how modern malware operated. One of the employees expressed that it was great that the company had such sophisticated controls but wanted to know how they could protect their personal equipment and accounts.

The question provided excellent context for my response. I responded that standard hygiene controls, such as patching and antivirus, are the most effective way to protect yourself. I then promised that our next presentation would be about how employees could defend themselves at home.

A few months later, the security team crafted a new presentation on how home users could protect themselves, including auto-updates, antivirus, browser controls, and how to freeze your credit. The presentation series was well-attended and became a hit with employees. With each presentation, we focused on the employee experience. We focused on a positive message with the assumption that if employees were more aware of the threats at home, they would also be more security aware at work and report events. Sure enough, security event reporting increased, and the firm's overall security improved.

⊕ Sharpshooter Tip

- Provide employees with actionable tips on how to protect their identities and assets. If they're more cautious when handling their personal assets, they will also be more aware and safeguard company assets.

↪ Misfire

- The security team gave a presentation that scared employees but failed to show them how to protect their personal equipment, leaving them feeling helpless.

INCIDENT RESPONSE

A common saying in the security industry is that there are two types of companies: companies that have been breached and companies that don't know they have been breached. Articulating this sentiment to senior executives provides fodder that most security teams are inept and security controls may not be worth the value. A better attitude is to state that your team is in a state of readiness and works to find and contain the intruders before they can cause significant harm. This section includes stories about incident response and the lessons learned along the way.

⮂ See something, say something

I had sponsored a "no-retaliation incident reporting" program within my company. Employees could report incidents, including clicking on a phishing attack, without fear of reprisal. Many upper management did not fully appreciate the no reprisal aspect and considered some employees stupid for "clicking on a link." As I presented the program to executive management, I demonstrated how most phishing attacks we received were designed to fool employees and that, inevitably, some attacks would get through. The key was to respond and contain the attack as quickly as possible. The company had a small call center for customer questions and concerns. The center had a website portal where customers could submit questions and upload documents. One of the call center representatives called the Information Security Team with something that seemed off: the representative had received a message through the portal from a customer. The message said, "Can you look at this?" and had a document attached. The representative tried to open the document, but nothing happened. That had been 2 hours ago. They remembered our security awareness training and decided to call my team. Is this document a corrupt file or something else?

Sharpshooter perspectives

SSP1: I'm a huge fan of a positive, safe culture for security awareness training, but I am confused by the inclusion of "clicking on a link" in your program. Usually, this results in remedial training and accountability for click-happy staff or those who were tricked. I agree that it is easy to be fooled, even as a security expert, and we want it to be safe, but how do you then measure the effectiveness of awareness training and reduced risk? If the program still includes click-through metrics and components with some form of an individual and corporate growth strategy to address those areas of weakness positively, I support this approach. However, if you are shooting for speed of response at the expense of awareness training, personal accountability, and metrics, I would not support that. In short, how do you define success in your social awareness program to reduce risk, meet compliance, and motivate employees to mature in their best practices?

SSP2: What controls are in place for vetting and processing documents shared in the portal? Are these documents scanned for malicious content, such as using a sandbox scanner to check for exploits in PDFs, etc.? It's conceivable that a "customer" might sign up specifically to target the company by uploading a hostile file, anticipating that the support team would open it. What segmentation controls and privileged access management (PAM) measures are in place for users? Additionally, what logical or physical separations exist to mitigate the portal's higher risk exposure, especially concerning external access to production environments and data lake storage? What level of trust should an employee have for documents within a portal compared to those received via external email? External email attachments should always be treated with suspicion until proven safe and reliable. I suspect the user didn't have the necessary software, such as Adobe Acrobat, to view the content uploaded to the portal. Depending on whether it was downloaded or rendered during the session, it was nonviewable within the browser or on their computer.

SSP3: I guess the file was malicious, and the local computer got malware because the representative clicked on the file. The representative did the right thing by contacting security. The system may have been designed to set the representative up for failure since there was likely no security scanning on documents submitted through the client portal. What was the representative supposed to do? I'm sure management expects representatives to address all customer concerns. How does the representative know when an attachment may be malicious? I fully support no retaliation incident reporting.

Target grouping

Two hours earlier, a call center representative received a document through the customer portal. When they clicked on it, the document refused to open. Fortunately, the representative contacted the Information Security Team to investigate.

The uploaded document turned out to be ransomware, slowly encrypting the files the representative had access to. During the incident response, the security team determined that no data had been transmitted from the infected workstation, and the infection was rapidly contained. The encrypted files were restored from the backup. We then provided additional training to help the call center employees recognize this type of attack.

⊕ Sharpshooter Tips

- At several companies I've worked at, I've heard, "We can't rely on users knowing what to click on and what not to!" This sentiment is usually followed by an argument to discontinue security awareness training or phishing tests. However, this incident illustrates the opposite: while the representative tried opening the file, their awareness training eventually kicked in, prompting them to contact the security team. Most companies still need robust security awareness training!
- Make reporting events nonretaliatory. I have had executive management ask me how some employees are "so stupid as to have clicked on that phishing email." Employees get overworked, sick, or sometimes mess up. By having a nonretaliatory policy, people will report events more readily, and in the end, it improves company security.
- Malware is constantly evolving, and eventually, something will get through. Have a solid incident response plan to investigate incidents and always have accessible backups.

⌐ Misfire

- If the security team had reviewed the customer portal, the lack of malware protection would have been discovered. Resist allowing any documents to come into the company unscanned.

⌇ Malicious insiders: We don't block porn?

It is not unusual to get incident alerts on a weekend. In the early 2010s, budget constraints prevented me from hiring an MSSP who could triage alerts on weekends. Fortunately, the number of alerts was manageable and could be handled by an on-call security analyst. The on-call analyst was sick this weekend, so I agreed to look at the alerts. On Sunday, around 2:00 PM, an alert came in from our network detection systems that someone had hit a drive-by-download website.

At the time, we didn't have a good endpoint detection and response (EDR) tool other than standard antivirus. Triaging the alert, I didn't see any antivirus alerts, so I examined the firewall traffic. As I worked through the traffic, I noticed something odd about the person who triggered the alerts. First, they were browsing a lot of pornographic sites. Five or six years ago, there had been a request from several executives at the firm not to block porn sites because the company might invest in the adult industry. This decision meant we blocked malware sites with URL filtering but not porn sites. I found another rule we had been too permissive about in the firewall traffic: we did not block sites that could be used to proxy web traffic.

As I examined the traffic, it became clear that the individual would go to an adult website, but since the site was infected with malware, it was blocked. Instead of heeding the block, they would go to a web proxy and hit the same website using the proxy site. As I continued, I was pleased to see the drive-by-download had been unsuccessful in compromising the laptop because it had been patched. I considered that they had loaned the laptop to a family member. I tried calling them on their cell phone, sent an email, and sent a pop-up message on their laptop. They never responded to my messages, and their computer went offline. As I reported the incident to the Chief Operating Officer (COO), the first words out of his mouth were "We don't block porn?"

Sharpshooter perspectives

SSP1: The response of the COO is an excellent opportunity to launch your sales pitch for the tools needed. I would explain the need to focus on reducing risk and blocking multiple threat vectors, not only porn – even though it is reasonable to expect that porn would be blocked due to the history of risk related to those sites. Hopefully, the discussion would provide approval to implement an advanced threat protection firewall and EDR solution that could work together to limit end-user risky behavior and protect the environment from many forms of malware.

SSP2: You didn't block porn? Proceed to HR immediately. Do not pass go. Do not collect $200. Seriously though, what was the business justification for allowing access to porn? Was it a company involved in the production of Adult Content? If it's acceptable, publish a dashboard that shows who is looking at what websites and allows it to be sorted and searched by category. Make it public, and watch your porn issue go limp.

Target grouping

Historically, the company blocked malware sites but not pornography websites. An employee uses a web proxy to circumvent the website blocks and access infected adult websites.

The reader is probably scratching their head on why the company did not block porn. Before I joined the company, the security team had attempted to block these websites, only to have the block removed by top-level executives. The justification was that the company might invest in companies in the adult entertainment business. Human Resources had not fought the issue, so they remained accessible.

I joined the company several years later, and although pornography sites were still unblocked, most of the top-level executives who had made the justification had shifted to other companies. The COO gave my team the authority to block the entire category immediately based on the incident, so we did, with no pushback. In addition, the COO fired the individual. Although pornography was not blocked, there were policies against circumventing security controls, providing ample justification for letting the individual go.

⊕ Sharpshooter Tips

- Establish a URL filtering firewall and gain approval to block specific categories.
- If there is a need to leave certain types of sites open, establish a group to restrict access.
- Review the categories that are open at least annually as business priorities change.

Misfire

- The security team should have reviewed the blocked website categories annually. Even with the justification for having adult sites open, the company should have restricted access based on business needs rather than having the category for everyone. In addition, web proxies should have been blocked.

Acquiring a security incident

The company I was working for had recently acquired a tiny company in the Asia Pacific region. The acquisition had fewer than ten employees, and the company had outsourced its IT support. The security team was contacted because upper management was asking to put the acquired company's servers onto our corporate network. The network and server teams were reluctant to do anything without a security review first.

Because upper management had escalated the request, I joined the due diligence calls, and soon, we spoke to the outsourced IT support team. As the call went on, we found that the company that we had acquired had an outsourced website and external-facing server that they accessed over a remote desktop (RDP). The external-facing server was used for research analysis and contained its customer management system. Upper management asked us to move the external-facing research server onto our corporate environment.

Performing due diligence with the outsourced IT support team, we inquired about their antivirus software. They named a well-known vendor and said they could show us the support console. As they opened the console, the IT person on the other side said, "That's odd. It looks like the anti-virus is offline." With that, he reactivated the antivirus, and alerts began to flow in like a stream of ants to a piece of discarded candy.

Sharpshooter perspectives

SSP1: There is no way I'm bringing these systems into my corporate network. We are talking about integrating a tiny organization, and the risk of potentially compromising our extensive corporate network is substantial. I'd focus on determining the solutions they need to perform their work, identify which ones are available in the corporate inventory, and then build the remaining ones. Data will be needed from the acquired company, so bring over only the required data, and make sure to use multiple antivirus engines or antimalware scanners before data ingestion.

SSP2: When it comes to mergers and adding unknown systems to the corporate network, I take an NTA position: Never Trust Anyone. If the security team doesn't give it a clean bill of health, it doesn't go on the network or into production. Think of it like getting a building permit. We deliver the requirements and inspect the system to ensure it complies. If all is good, you get an Authority to Operate (ATO) and can move it into production. If you don't pass the inspection, there is no ATO, and we give you a list of issues to correct before reinspection; you can't move it until it passes and the ATO is granted.

Target grouping

The security team had been asked to allow servers from a recent acquisition onto the corporate network. The servers were shown to have the antivirus turned off. When it was turned on, it was clear that the servers were infected.

The next few weeks were difficult because the outsourced IT had no information security personnel and were unaccustomed to performing incident response. Because of their lack of resources, my incident response team worked with them to collect and analyze evidence. It was discovered that someone using an IP address in Singapore had used a brute-force password-guessing tool to break into the server's administrator account over remote desktop protocol. After obtaining access, they opened a browser and navigated to a website in the Ukraine that hosted a drive-by-download that compromised and installed a bot onto the server. Using the browser exploit was odd because they already had complete control of the server with the administrator account.

Speaking with the outsourced IT team, I learned they had accidentally disabled the antivirus months earlier and had not noticed it was off. Throughout the process, I kept the executives updated on the investigation. About 5 days into it, I got some surprising news: the president of the acquired company had sent a notice to all their customers about the compromise without consulting anyone in our company. Fortunately, the customers took the news reasonably well.

Ultimately, the customer management system was rebuilt from trusted software sources, and the data was moved to a new set of servers.

⊕ Sharpshooter Tip

- Before letting new systems onto the network, a security review should be performed. In this case, several mistakes led to a more extensive compromise. In addition, the security team should be brought in to conduct due diligence on new acquisitions.

⌐⁓ Misfire

- The security team should have been brought into the acquisition process, where the compromise may have been detected earlier. In addition, the acquiring company could have used the security issues as leverage because the compromise could reduce the new company's value. Second, the acquisition should have been required to follow the leading company's incident response and notification procedures, providing consistent reporting to clients.

≥ Trust but verify

I had recently been hired as the CISO of a technology company, a subsidiary of a large manufacturing company. Just before I joined the company, the Information Security Team had gone through a reorganization and was still learning to work together. Concerned about the technology company's IP, the CEO hired a well-known penetration testing company to perform a penetration test to determine if breaking into the technology subsidiary from the manufacturing side was possible.

The lead penetration tester was one of the best I had worked with, and during testing, I maintained good communication with him and the team to follow their progress. Because our main adversaries were nation-states,

I wanted the penetration tester to progress as far as possible and discover as much as possible. As they progressed, I provided them with some bread-crumbs to get them closer to breaking into the technology company's net-work. During one of the sessions, the tester off-handedly said, "Have you checked for this vulnerability on your load balancers? It was just announced and is remotely exploitable." I heard of the vulnerability and sent it to the teams that patched the IT systems. Later that day, the IT team responsible for the load balancer reported that it was not vulnerable. The penetration testers could not break into our network even with the breadcrumbs.

We worked through the report findings, which were all low-risk. Several weeks later, an architect on my team asked me if I had heard about a vulner-ability in the load balancers. I replied that I had and that IT had said we were not vulnerable. "That's odd," the architect stated, "I just ran a scan, and it says that we are vulnerable."

Sharpshooter perspectives

SSP1: I've seen two cases where previously verified patched vulnerabilities resurface. The first is where additional patches are applied, and it reopens previous vulnerabilities. This wasn't that uncommon 10–20 years ago. So, a vulnerability scan had to be completed after applying a patch to ensure the patch was successful and that previously patched vulnerabilities didn't resur-face. The second is when a restore is used, and the admin doesn't take time to complete the required patch sequence to obtain concurrency. I'd start the investigation by looking at those two situations since it seems unlikely that IT told you incorrectly and the auditors didn't find it.

SSP2: Just because someone is speaking doesn't mean they know what they're talking about or telling the truth. Sad but real. Sometimes, folks say things because they need to respond and have an ego/identity issue with being the person who "knows" things, so they are uncomfortable letting others know they don't know a specific thing. Honest mistakes happen, but when it comes to things of this nature, it is always better to have two sets of eyes review it versus one. It is also crucial to ensure a culture of safety and transparency that allows for mistakes and "non-perfection" among the staff so ego-driven communication is minimized.

Target grouping

After being alerted of a potential vulnerability in our load balancers, the Information Security Team alerted the IT team in charge of patching. The IT team still thought the load balancer was not vulnerable, but they were wrong; later testing revealed that the load balancers were vulnerable.

Unfortunately, the IT team misread the original vulnerability report and interpreted that we were not vulnerable. We investigated the load balancers and discovered they had been compromised multiple times. The load balancers had a webshell installed, which gave the attackers access to the system. In addition, numerous automated jobs were installed in the task schedulers.

I was still relatively new at the company and pulled aside my SOC manager to work through the incident. I had yet to work with him on an incident and asked him about his incident response plans. Unfortunately, the SOC team had never drafted an incident response plan and needed to know where to begin. Based on that, I immediately contracted an external incident response company to come in and analyze the systems. The external team was worth the cost and could determine that the attackers had not moved beyond the compromised load balancers. Based on the analysis, I asked the SOC team to decide which traffic could have been exposed through the load balancer. After a few weeks, I asked the SOC manager for a status report. He was still working with the IT group to generate a list of traffic that went through the load balancers. After another couple of weeks, I pressed IT and the SOC Manager to see what traffic passed through the load balances. As it turned out, the load balancers routed email from our external gateways to the internal exchange servers. In addition, some financial data was loaded from a corporate system to our systems.

Some companies use load balancers to encrypt the traffic between external sources and internal servers. I had the SOC team determine which data was encrypted from each originating system to the system on the other end and if any decryption was performed at the load balancer. As it turned out, the traffic was being encrypted at the originating system and was encrypted as it passed through the load balancers. It was also not being decrypted at the load balancers. Using this information and the report from the incident response provider, we could file an incident report with our customers. In the end, we decided to rebuild the load balancers.

Misfire

- For this story, it's better to examine the misfires first. The company recently underwent a complete information technology and security reorganization. Before restructuring, the company viewed security as a dysfunctional area. In the reorganization, the company retained several individuals who did not have the skill or desire to carry out their roles. In addition, the company didn't provide much support to the IT and Security teams, making it a disheartening area to work in. It usually takes a new CISO about 6 months to create an accounting of what is working and what is not. Unfortunately, an incident struck before I could develop a roadmap. The missing elements were:
  1. No established external scanning program to identify vulnerabilities.
  2. Over-reliance on the IT organization to understand the vulnerabilities in the environment.
  3. Lack of visibility on network traffic entering or leaving the environment.

4. An incident response team without a response plan, training, or tools.
5. No established contract with an external incident response provider.

⊕ Sharpshooter Tips

- As a new CISO, concentrate on attack surface management as you enter the organization. Does your company have processes to scan and patch vulnerabilities? In addition, incident response capabilities should be established before an event. Work with the team to develop incident response tools, procedures, and security vendors on retainer before the incident strikes.

⊇ Website vulnerabilities: Blacklisted!

My firm was merging with another company in Europe. As part of the due diligence, I learned about different IT systems and worked closely with the IT teams to understand how each system would be integrated. Understandably, we had two different websites. The integration team rebranded both sites with the new company logo but kept the two sites separate since the two companies had different products.

During the due diligence, I noticed that the European website had a form that someone could fill in an email address and send a product description to that email address. In looking at the website, I also noted that the form did not have a CAPTCHA (Completely Automated Public Turing test to tell Computers and Humans Apart), which would reduce the ability for someone to automate an attack against the form.

I followed up with the VP of Development, who informed me there had never been issues with anyone abusing the form. I asked about the architecture, and he responded that an email from the form was sent to our Exchange servers for processing. I explained to him that someone could use the form to automate an attack, which would come from our Exchange servers and could get us on email blacklists. "Well," he replied, "It hasn't happened yet."

Sharpshooter perspectives

SSP1: President Thomas Jefferson said, "If ignorance is bliss, why aren't more people happy." Risk acceptance comes in many forms, including denial and ignorance. I call this the "Ostrich Syndrome." Leaders sometimes stick their heads in the sand – in fact, shout it out and are proud of it – intentionally ignoring risk. They are happy to adopt the attitude, "If it ain't broke, don't fix it," while ignoring that it is incomplete and highly vulnerable to attack. Perhaps the executive would care more if the website impact was further qualified with additional outcomes. For example, blacklisting by others could disrupt use of the website for up to 30 days or longer. Additionally, automated attacks against the back-end server, where input controls are lacking, could lead to exploitation and other serious issues. A complete picture of the risk, such as recent security incidents with other companies, may be enough to change the hearts and minds of the Ostriches that live among us.

SSP2: I'm sure the VP of Development focuses solely on managing roaring fires, likely because he has many to deal with, making him unwilling to address problems that aren't immediately critical. In this situation, I'm looking for a technical solution demonstrating the need for a permanent fix. If someone automates an attack using the form, I can use the evidence to educate the VP and advocate for a long-term solution.

The technical solution I propose involves working with the Exchange Admin to implement message rate limits, throttling, and alerting. These measures will help detect and mitigate automated attacks on the form, providing concrete evidence to justify the necessary permanent changes.

Target grouping

One of our company websites had no rate-limiting protection, such as a CAPTCHA, which would allow an attacker to use a contact form to spam another email account. This could get our email servers on a blacklist.

A few months later, an automation attack was launched against the form, which caused a DoS attack on another mail server. The email team noticed it because the messages were bouncing back. The VP of Development put in a block because only a single IP was being used. Because we were going through a merger, the development team was used to operating independently and notified the security team several days after the attack.

I reiterated that a CAPTCHA or other form of rate limiting should be put in place and reiterated the risk of our company being put on a blacklist. Blacklisting had the potential to knock our email off most of the internet. The VP noted that the attack had happened only once and that there were plans to rework the website code as part of the merger. Again, he accepted the risk, avoiding any effort spent on the old site.

Four months later, another automated attack used the contact form to perform a DoS against a Chinese social media site. With the attack, we were put on an email blacklist in Europe. Because of the merger, mail was also routed through mail servers in the United States, which were not blacklisted, but the servers in Europe lost the ability to send email to companies using that blacklist. It was a painful 8 hours to get off the blacklist. After the incident, the VP of Development finally approved rate limiting for the form.

⊕ Sharpshooter Tips

- Blacklisting like this can be prevented using several secure development practices:
  - If possible, have the development team use an API to send external email through a trusted email service and block the ability to send email using the company email servers.

- Depending on the type of website form, implement a rate-limiting mechanism on the backend, such as a CAPTCHA or email throttling.
- Set up secure code training with the development team so they know common attacks and how to prevent them.

🐾 Misfire

- Because this was a merger, specific teams ignored security to meet their objectives. We should have worked with executive management earlier to set up a risk acceptance process to help everyone understand the potential consequences.

2 Chasing our tails

I was facing one of my most difficult challenges as a CISO. Hours ago, an endpoint agent on one of our internal servers detected a copy of mimikatz, an adversarial tool that extracts hashes, pins, and passwords from running servers. The endpoint generated an alert and sent it to the SOC team, who escalated it to me. Finding mimikatz on an internal server meant someone was in the network and had already been moving across our environment. Not good!

The endpoint tool quarantined the file, and the SOC team was able to pull the file out of quarantine to perform malware analysis. The file was, without a doubt, mimikatz. The attacker had not even bothered to rename the file and left it named mimikatz. As the SOC team continued their work, they discovered several things:

1. The account that last modified the file was now a security identifier (SID), a string of numbers left in place after the active directory account was removed.
2. We immediately pulled 6 months of server logs, but none of them revealed anything about the file or the account that had left it.
3. In checking the last modifed date, it suggested that mimikatz had been on the server for at least 18 months!

Piecing all of this together suggested that our attacker had been in the environment for a long time, and the tracks had grown cold.

Sharpshooter perspectives

SSP1: Ouch, this is very troubling, and my first thought as a CISO is that "I'm compromised." This information requires immediate activation of the Incident Response Plan (IRP). I've learned that while this is a serious concern and requires immediate attention, it is better to take a more systematic approach and plan remediation if they've been in your environment that long. If they have been in the environment that long, clean-up has to be well-planned and orchestrated. Also, out-of-band communications should be used for the incident so they don't know they've been detected. I can't wait to hear the conclusion of your investigation and what remediation steps were required. The title of this story makes me think there is a twist here.

SSP2: This is bad! This is very bad! Even if it can be explained, it is still very, very bad. How bad, you might ask? If the compromise truly spans 18+ months, it could be an "end of an organization" level compromise. On the other hand, it could be a "run-of-the-mill" bad if an unauthorized tool was used for legitimate purposes. Either way, it demands immediate attention and appropriate action.

Target grouping

A tool used to steal password hashes was discovered on a company server, but the access logs only went back 6 months. The incident was difficult to investigate because of the time that had passed. There were a couple of odd facts that came out of the investigation: (1) mimikatz had not been renamed, and (2) the dates began to line up with a penetration test that had taken place over 18 months ago.

The incident response team followed the IRP, still assuming that it was a true compromise. Fortunately, backup tapes allowed the team to pull the original mimikatz file and gain information on who the SID belonged to. It was linked to a penetration tester who conducted the test over 18 months ago. The server logs were also restored, and the information in the SID was corroborated. The penetration tester had used their account to gain access to the server and then used mimikatz to steal additional credentials. Although unplanned, the exercise taught me and the team valuable lessons.

⊕ Sharpshooter Tips

- Cybersecurity breaches are a common source of fear and anxiety for CISOs. Understanding the threat, having the right people, processes, and tools in place, and understanding that you are not alone are keys to breach readiness and mental preparedness.
- Having experienced CISOs who have faced and overcome breaches or failures is essential for companies because these individuals possess the resilience and knowledge necessary to handle and recover from incidents effectively.
- Internal and external audit groups can cause anxiety and frustration. Having a well-managed and definable program is the antidote to audit overreach.
- Having actionable backups was crucial for obtaining logs on the incident. Confirm how much your backups capture and test recovery.

- Periodically running antimalware scans across the environment should be standard practice. This malicious tool would have been flagged much earlier.
- Add a clause in the contract requiring penetration testers to clean up any tools left behind in the environment. As a CISO, you do not want attackers to find and exploit readily available tools to compromise your environment.

Misfire

- The company should have been running antimalware scans more regularly across the environment, which would have detected the event earlier. In addition, penetration testers should clean up after the test is over.

Chapter 5

Disaster readiness planning

When catastrophe strikes, preparation separates the panicked from the poised. As such, CISOs must anticipate and mitigate potential crises through strategic foresight. Building a resilient, flexible infrastructure ensures stability amidst challenges, not from fear but informed risk management. Vigilance and realism are crucial in detecting and responding to threats, emphasizing the importance of internal strength and discipline in readiness efforts. By fostering a culture of preparedness, the CISO can help transform panic into action, ensuring survival and resilience in the face of adversity. This chapter outlines the essentials of proactive disaster readiness that will be crucial to your success when the inevitable happens.

BUSINESS CONTINUITY

⮂ Plan for the worst – hope for the best

As the IT Incident Response Leader for a $10B international engineering and construction company, I was invited to the company-wide business impact analysis (BIA) planning meeting as the IT representative. This was a well-planned and orchestrated meeting where representatives from across the company evaluated the potential impact of various disruptions on each business function, considering factors like downtime, financial loss, customer impact, legal and regulatory consequences, and reputational damage. Once the disruptions were identified, the team defined recovery objectives and mitigation plans for each disruption, which were documented and sent to the division executive vice presidents and the president for approval. This exercise was immediately after Operation Iraqi Freedom in 2003, and we had contracts in Iraq associated with the reconstruction effort. There were many fascinating scenarios that the team discussed and planned for, but there was one that is still very vivid after 20 years. What are we going to do if we have an employee in Iraq who is lost due to an untimely death? This is something that I had never thought to develop for an IT security incident. While it was much broader than IT, and the humanity side had to be addressed, it was still applicable since we provided computer and network services within the green

DOI: 10.1201/9781032720500-6

zone. There was the potential for availability interruption. How should we plan for this tragedy, and how would you prioritize the response?

Sharpshooter perspectives (SSP)

SSP1: This is an excellent opportunity to look at critical person risk. This can occur when someone is either injured or is out on long-term disability. It also can happen when a spouse or relative is the affected individual, and the employee needs to take time off to assist with the associated trauma. As you mentioned, there are two aspects to this: (1) operationally, how does the company deal with the loss of the individual? and (2) how does everyone emotionally deal with the loss? We have all developed key person risk charts to assess our skills and ability to cover specific procedures. Still, key person risk charts can only take an understanding of how operationally things will be handled so far.

SSP2: This reminds me of a story. Before I was a CISO, I was a developer and performed release management for the group. It was a small group, and the question arose who would take over releases if I was sick or taking a vacation. We had several developers and project managers, but I had taken over the code repository and servers from another developer who had left the company. Management had me train the quality assurance person to move tagged code from the repository into production. I also documented the process using a standard operating procedure (SOP). An issue came up when I was sick with the flu. I felt better in the afternoon and logged in to see how the SOP was working. My backup had sent everyone a message asking to postpone releasing code that day since I was out and that I would deploy it tomorrow. This situation taught me that more than assessing critical person risk is needed. An SOP is not enough, either. What is required is cross-training of backups, which can be challenging for a small team.

SSP3: A grim reminder that none of us are getting out of this alive. This is an excellent example of identifying critical constraints in your program and organization. As part of a quarterly review of our program and staff, we review every employee on the team to understand if they are "one deep" and if there is a backup for them. We also use this time to understand who may or may not be a flight risk. When identifying the "one deep with flight risk," we immediately focus on ensuring a backup plan. It is also a good reminder that when only one individual can perform a specific task, you are automatically rate-limited as to how much of that capability you can provide to the organization. Another factor to consider is the training and resilience of talent profiles.

Target grouping

The worst-case scenario did happen, and while tragic, the company was well prepared to deal with the situation and help meet each family's needs. The personal impact it had on me was that the company President was a person of great moral character. Seeing these acts of humanity told me that I was in the right company and wanted to do everything in my power to help the President and the company succeed.

The team decided that preparation would include establishing the correct State Department contacts to help manage the situation and identifying two Family Ambassadors that would be dispatched immediately to be with or near the family should the unthinkable occur. These ambassadors had to be highly passionate and caring people with skills to make things happen during chaos. Each would be authorized to spend $200,000 with no questions asked to do whatever the family needed. Items discussed for expenditures were plane tickets for family members, hotels, funeral expenses, meals, and clothes. Should the $200,000 not be enough, they would speak with the responsible Executive VP, who would approve the additional expenditure as a control formality. I remember the company President stating, "No one in our work family is going to need anything in this situation if I can help it." Unfortunately, two employees were killed in Iraq in separate incidents on the same day within 2 months of this planning exercise. The Family Ambassadors were immediately dispatched and ensured that the families' needs were met. They stayed well after the funerals to help the widows with all the overwhelming legal and financial matters. While this is a tragic story, it shows that BIA planning is essential. I've told this story before, and some have commented that the business was trying to protect itself. While someone might interpret that what the BIA team planned was the best thing for the company, that was never part of the discussion. It was all about displaying humanity, doing the right thing, and caring for our people.

⊕ Sharpshooter Tip

- Doing the right thing for people as individuals should always be a higher priority than any business or security objective. When doing the right thing, don't worry about perceptions. You can't control them anyway. In the end, it is the people and their well-being that matter. If you work for a company where that isn't the case, you are in the wrong place, so hit the eject button, and your parachute will guide you somewhere that deserves someone of your high character.

⌐ Misfires

- The real misfire with BIA planning is not doing it. Whether you can spend 1 hour or 3 days with BIA planning, you will be in a better preparation state than having done no planning. Will you ever plan for every scenario? Probably not, but with some dedicated effort and support from key stakeholders in the company, you can achieve well over 80%.
- Planning for a disaster is incomplete without testing the plan. Testing identifies gaps and weaknesses that theoretical planning might miss.

Additionally, testing helps refine communication and resource allocation strategies that often fail under actual disaster conditions.

੨ It's only a disaster if you can't recover

Every morning at 0500, the backup system needed to have the removable media changed and placed in the box that left the facility for offsite storage. In the pre-dawn hours of every morning, this had happened without fail for the last 5 years. As the team's new manager responsible for the backup and restore of primary systems in the enterprise data center, it fell on me to ensure that the process was performed when the employee on that shift was on vacation or out. On one of these uncommon early mornings, I started to do the media change and noticed the age of the specific media being swapped out. It was about 5 years old. The specific lifecycle for these was only 3 years, so I was interested in how these were still in service. Upon further checking, I found out that all of the media was about the same age and hadn't been updated since the implementation of the original solution.

The systems were on a managed lifecycle and had been updated 2 years prior, but for some reason, the media currently in use appeared to be from the initial purchase. That would mean the media had been used over and over without notice or attention to potential expiration dates. I also checked the last time a full restore test had been done and was surprised to see it was over 24 months prior. Now I was worried. If anything happened, could we actually recover, or were we just in the final act of performing backup theater?

Sharpshooter perspectives

SSP1: In 2016, I was tasked with recovering source code from a set of 3.5-inch floppies. Fortunately, I was asked to perform the work to see if the source code could be recovered rather than in an actual disaster recovery (DR) scenario. It took a long time. I had to find a USB floppy drive and then get the media to mount properly. The disks were old, and I had trouble mounting on the floppy drive. I worked on it and finally recovered the source code. The lesson I learned is that media becomes more difficult to work with the older it gets; if what you are backing up is important, you have to keep up with the technology, or recovery of the backups can take a long time if you can even get the media to work at all!

SSP2: It's great that they had you as the new manager and that you were on top of things. It's time to reach into the emergency fund and buy new tapes now! It's concerning that the tape systems were in lifecycle management but not the tapes. I'd also perform a full restore once the new tapes are in place. I worked for an organization that went to new tapes, and when we did a test restore, we learned that every tenth tape was bad. That would have been a bad day if we discovered that during an emergency restore.

Target grouping

You can imagine how quickly I purchased and put a new set of tapes into service. In addition to that activity, we did a test restore from the tapes in rotation and found that only one of the six backup sets in rotation was actually valid. One in six! Seriously?! That is less than 17%, which meant, from a practical standpoint, that the organization's last effective backup was over a month old at the time of the test. Organizationally, we had been working under a false sense of security and would have had to take a month's worth of data loss. A non-trivial impact for a ~$6B organization. All portions of the backup and recovery solution were put under service lifecycle management, and recovery testing was moved to quarterly.

⊕ Sharpshooter Tips

- A process can be performed without error and still not be functional. This process seemed to be working, but it wasn't. Regular testing of a solution's validity is required to ensure its effectiveness will be there when you need it.
- If you see something, say something. If something doesn't look right, then it probably isn't.
- It might be a good idea to temporarily rotate various people into different roles to have a "fresh set of eyes" to review repetitive activities and ensure processes perform as they should.

Misfire

- Assuming that just because a process hadn't failed, it is still functioning as intended. The tape backup process was working flawlessly. However, the ability to use those backups to restore had failed in silence behind the scenes.

IMPACT ASSESSMENTS

⊇ Can't see the forest for the trees

I am reminded of a time when I was brought in to improve the company's technical risk position before its next annual security review. This organization had failed several audits and experienced a material breach the previous year. Taking a lead security role in a company that saw its defenses crumble the last year is not for the faint of heart. The quick exit of the former CISO amid a collection of security concerns left a clear message: what was currently being done wasn't working and needed to change.

The company was well regarded, but beneath the surface, its IT infrastructure was a patchwork quilt of vulnerabilities. My hiring was met with skepticism and hope from the executive team and the cybersecurity department.

The obvious question on the table was: How could I succeed where others had failed?

The first few weeks were a whirlwind of meetings, reports, and assessments. I dove into the records of previous years' security failures to understand the root causes. It didn't take long to uncover the first major issue: a glaring lack of understanding of the company's core assets. The "crown jewels," as they are often referred to in cybersecurity parlance, were not clearly identified, leaving the company's most valuable information under-protected.

As I delved deeper, another revelation came to light. The majority of the cybersecurity budget and effort were channeled into protecting systems and data that, while important, were not critical to the company's survival or competitive edge.

Sharpshooter perspectives

SSP1: The classic cybersecurity story – follow the data and the money, and hopefully, the money is following the critical data. Great job on your part for identifying that was not the case. It sounds like a case where you may have to tear down the complete protection strategy and rebuild from the ground up. Create your plan, communicate it clearly, secure buy-in from key stake-holders, educate those previously involved and who missed the critical assets, implement the plan, and measure the effectiveness of its implementation.

SSP2: Moving from symptoms to the root cause is always a Sharpshooter's objective. The author has excellent insight and experience as a leader to real-ize and listen to the company's history, leading to identifying current chal-lenges and core needs. A transformation is needed, but is it cultural, technical, procedural, operational, or all of the above?

I heard the term "patchwork" in relation to threat and vulnerability manage-ment (TVM), which I interpreted as fragmented, incomplete, and indicative of an immature TVM lifecycle – a situation all too common in many organiza-tions. This patchwork often leads to higher risk exposure throughout the TVM lifecycle, with vulnerabilities and nested exposures remaining unseen and unmanaged for extended periods, potentially resulting in breaches over time.

I also noticed a lack of understanding of identifying "crown jewels." Gathering stakeholders for a 1- to 2-hour session can determine these critical assets quickly. During this meeting, we use a whiteboard to:

1. List our high-value assets (no wrong answers here).
2. Discuss how these assets are protected.
3. Evaluate the strength or weakness of defenses for each asset.

We then prioritize these assets, discussing the "why" factor related to risk, clarifying impact and likelihood, and assessing the specific risks to the com-pany. This exercise often leads to spontaneous discussions around key risk areas such as operational downtime, breach/disclosure/reputational risks, and IP loss.

This process is an excellent starting point for fostering cultural change, enhancing transparency, and shifting the focus to risk management rather than merely addressing patching and security engineering issues.

Lastly, you need senior management that balances tactical, operational, and strategic leadership. This plan must include a roadmap and ongoing maintenance and support. Your competitive business plan is evident as part of this, with security, brand, and other elements as core components.

SSP3: I was put into a similar situation as a longtime Information Security Manager. Management hired a new CISO, who started in the right direction by stating that they could prioritize risk management by working to protect the company's critical assets. I raised my hand and said no one has defined a list of critical assets, but I can help put it together. The CISO ignored me because I was lower on the management chain. He said the DBAs would know where the critical data was and then tasked them with compiling the list. The DBAs then delegated the task back to me.

I ended up creating the list but received little credit! This experience taught me the importance of listening to different teams over the years and gathering information about the types of data stored in each system. Sometimes, the crusty technology person who has been with the company for a while can provide valuable insights into what each system contains.

Target grouping

We spend a lot of time working on "how" to protect our organizations, but we often spend less time than we should on the "what" or "why" of that equation. In this instance, it was a classic case of misdirected priorities, which started to explain the ease with which cyber attackers had previously penetrated the company's defenses.

I proposed conducting a BIA focused on security and privacy. This assessment would help identify the core assets and provide a clear picture of the potential impacts of various cyber threats. It was an approach that hadn't been tried before in the company because there was a false belief that "business" and "security" were somehow separate. While some executives were hesitant, the memory of the previous year's failures led them to give the green light.

The BIA process was thorough and eye-opening, and the results were shocking yet unsurprising: roughly 80% of the cybersecurity program's resources were focused on protecting the 20% of essentially non-critical assets. The actual "crown jewels" were left almost bare, exposed to any threat actor savvy enough to spot the opportunity.

Using the BIA, we could paint a clear picture of resource misallocation, and the executive team could see the gaps in the organization's defense as plainly as holes in a block of Swiss cheese.

The solution was as elegant as it was effective: apply the Pareto Principle (80/20 rule) by reallocating resources to ensure that most of the cybersecurity budget and efforts were dedicated to protecting the "crown jewels." It was a strategic pivot that required reshuffling priorities, reallocating funds, and, most importantly, changing the company culture around cybersecurity.

Implementing the new strategy was a massive undertaking. Old security protocols with years of entrenchment were overhauled, and new defenses were put in place with the correct focus. Training programs were revamped to ensure that every team member was aligned with the latest priorities, and some asset-specific solutions were deployed to strengthen the security posture of hyper-critical assets.

The transformation was not immediate, but the results began to show over the next few months. The operational security was at levels the company had never seen before, and the risk of a breach had been drastically reduced. All of this was achieved with almost no change to the cybersecurity budget – just a more intelligent allocation of resources.

The following assessment was a night and day difference from the previous year. By understanding the value of the company assets and protecting them with a focused, data-driven approach, we secured the digital present and paved the way for a safer future.

⊕ Sharpshooter Tip

- Identify which assets need protection and start there. In legacy organizations, it's often about understanding the current identity and priorities of the organization, which may differ significantly from the past. What worked to get you here won't necessarily keep you or take you where you want to go.

⌐ Misfire

- Organizations often get bogged down by fighting daily fires caused by not doing the fundamentals well enough. Doing the hard thing first is generally good advice, but knowing why you are doing the hard thing is more important.

⧠ Friends of a friend – hidden third-party risks

Have you ever had that friend who had a friend you would never have as a friend, but somehow, you ended up having to deal with the negatives of that friend? Well, that is how I describe the kind of third-party risks we had to deal with once while in the middle of a merger and acquisition. Our organization was acquiring another site when a third party experienced a cybersecurity incident. We had no idea the two companies were connected until the incident occurred.

Here is where things get interesting:

- It wasn't the third party that notified us of the event.
- It wasn't the organization that we were acquiring.

- It wasn't our SOC.
- It was people who showed up with badges.

Sharpshooter perspectives

SSP1: A direct network connection with anyone is generally a bad idea. I have seen a few exceptions that worked well through careful planning, security measures, and contractual agreements. During the lesson-learned portion of the incident, I'd ask why there was a third-party network connection. Is there a valid business reason? I thought there may have been a business need at one time, but that need had long passed, and no one bothered to take down the link.

SSP2: I have seen this problem through a lot of mechanisms:

1. IoT devices that people forget about. I have seen ATMs and credit card processing machines get compromised,
2. "Vendor access procedures" that only the application owners know about, and
3. Tracking down connections to identify all the direct lines and site-to-site VPNs. Getting a good account of all those connections is hard without a good relationship with the network team.

SSP3: Third-party risk management, extending to 4th and 5th parties and beyond, is challenging in a world of diverse, integrated shared responsibilities and complex infrastructures with interdependencies.

A key strategy for managing this is prioritizing technology and vendor onboarding to align with risk management priorities. This is especially crucial for architectures tied to high-value assets, identity and access management (IAM), and understanding where data resides and how it is protected throughout the operations lifecycle for all technologies and processes. This is no small feat, but it is necessary as we introduce change in the future. Ensuring ownership is crucial as we shift responsibilities to various external agencies while managing our high-value assets and partnered interdependencies effectively, avoiding unexpected issues such as badge surprises.

Target grouping

After a bit of investigation to determine that the badges were legitimate, we determined that attackers gained access to the network of a third- party partner who happened to have a direct connection between that organization and the one we were merging with. This connection was unknown to us, as it was not documented in any network architecture diagrams, and activity across it was sporadic. Needless to say, that connection was severed immediately and with extreme prejudice.

The next few days entailed an immediate lockdown and a remediation sprint to address high-priority vulnerabilities. Working closely with the third-party company to understand the specifics, we obtained all related indicators of compromise (IoC) and performed exploit-specific discovery and proactive protection. Luckily, the minimal controls and ACLs on the switch and the connection had proved enough to protect us until we could adopt a more appropriate security posture.

During the investigation, we discovered that this network connection was not intended as a long-term solution when it was put in place and needed to be more secure than we would have liked to see. We also discovered that our network architecture documentation needed to be updated and complete. This led to some immediate and focused work that can be considered the silver lining of this event. Pending security work was immediately accelerated and implemented, creating a better security posture; the incorrect network architecture discovered was corrected, and a business continuity plan (BCP) was developed and implemented for future use.

The primary lesson learned is that trust isn't enough; you must verify. You never trust and always verify when connecting with third-party environments and making new friends. Because it isn't just your new friend you have to worry about; it's everyone else who is "friends" with them who also plays into your risk posture.

This cybersecurity incident also reminded us that third-party breaches can significantly threaten organizations; sometimes, third parties are fourth, fifth, or beyond. It is essential to have a strong understanding of the security posture of your third-party vendors as far down the chain as you can achieve and to have a plan in place for responding to a breach anywhere along that chain.

It is also essential to have up-to-date and complete network architecture documentation. This documentation should help identify and mitigate potential vulnerabilities before they happen. Good documentation is the alpha and omega of any security program. It can be a critical factor in a successful or unsuccessful incident response.

⊕ Sharpshooter Tips

- M&A activity is always tricky, but don't rely on your partners' attestations. Use that as a starting point whenever possible and do your own investigations and discovery. It may seem like extra work, but it is essential for everyone's sake.
- Trust is not just a two-way street. It is a spaghetti mess of relationships, past and present. The longer an organization has been around, the more complex the environment.

⌐ Misfires

- The need for up-to-date and complete network architecture documentation had unacceptably exposed the organization.
- Trusting in the completeness of the documentation we had been given. Saving time by not validating costs much more in lost productivity and emergency work after the fact.

- A friend of a friend can ruin your party as quickly as the friend you know can. Apply the same level of scrutiny to any possible connections back to you. Test all links in the chains that connect back.
- Keep your documentation up to date and relevant to the environment. A configuration change requires a change in documentation of the current state. Don't close the ticket until that part is done because the work isn't complete.
- Know who your state and federal partners are before you need to know them. The time saved in validating threats or getting assistance can distinguish between a positive and negative outcome.

TABLETOP EXERCISES

♻ That's not my job

One consistent area for improvement that I find when beginning new CISO roles is the absence of a BCP. Not the lack of a comprehensive plan – the lack of any plan. In taking a new role, I once again found no BCP. Even though some incidents and sporadic notes had been assembled with some deficiencies noted, no corrective actions were identified, nor was there any evidence of action taken. The leaders in the company had identified that they needed lists of whom to contact if an incident occurred but had no clue what a BCP should contain. Since there was no apparent position where the BCP responsibility fell, I took it upon myself. After identifying what I considered the major stakeholders and interviewing them about preparations for business-impacting incidents, I determined that it would be most beneficial to use my years of experience in business continuity planning to write the first draft of the BCP instead of trying to educate the whole company and going through the normal three-step process of:

1. Conducting a BIA.
2. Develop a business-focused risk assessment.
3. Develop the BCP to address the most significant business impact and highest risks.

Doing so would take a very long time, and having a plan that could be used, even with deficiencies, would be better than the current situation. Also, the company leadership would respond better to seeing a product that is relatable to them, and in other words, it would be easier for them to throw darts at a dart board than to have original thoughts. I completed what I thought to be a 90% draft. I routed it for stakeholder comments, writing a specific email to each stakeholder explaining why they were a stakeholder and what areas of the document they were most critical to review. A month goes by, and crickets. Then, an incident occurred in the building, and the head of the company contacted me. I said, "The steps to take are on page 8 of the BCP." He quickly appointed me to address the incident. Fortunately, but unfortunately, the incident was so minor that it and the BCP were soon forgotten. No worries, the next step was to hold a tabletop BCP exercise, and a scenario was developed that would most likely gain and keep their interest. The exercise was scheduled with the identified stakeholders, who were recommended to review the BCP and their specific roles.

The BCP exercise was well attended. Initially, it was a little awkward, as almost no one had reviewed the document. We got about 15 minutes into the exercise when the Incident Response Team (IRT) Leader asked one of the participants to identify what they would do in response to an item. The response was, "That's not my job?" and the IRT Leader retorted, "The document says it's your job!" Had I misfired? Was I making progress?

Sharpshooter perspectives

SSP1: This is a hot button for me. If a job needs to be done and it truly doesn't have an owner, it doesn't matter if it is yours or not. Just do it. That is the right thing for the more significant situation. However, that doesn't mean you need to own it. It means you get to create it and, to the point in the story, give the organization something initial to work with and mature. Part of that creation work is deciding where it needs to be in the organization and who the ultimate owner needs to be. Create it and delegate it. If they complain, then at least the conversion is started, and the revisions can begin. Progress is being made because the conversation is happening if nothing else.

SSP2: The tone from the top is essential in changing this aspect. Hopefully, this exchange ended amicably. From my experience, the events that typically change the tone at the top usually involve actual incidents. I was in New Jersey during 9/11 and was across the river from New York City that day. It was a horrible day. However, executives at my company took it seriously and set up a BCP program shortly afterward. I can contrast that to another company I worked for, where the President told me that BCP was too time-consuming and that we would not put in anything beyond DR plans. Having had to work through several incidents, the IT teams hated not having BCP plans because the business always assumed that IT could drop everything and fix any outage. Communicating with upper management and finding buy-in is critical to driving accountability.

Target grouping

The discussion thus far had not been a misfire. This dialogue began a healthy debate among the senior management, saying that everyone needs to understand their role during an incident and that their role in the company impacts them during an incident. Still, it doesn't need to be the same. Their role during an incident should be what is defined in the BCP. We made it through the 2-hour exercise, and wow, did I receive comments on the BCP after the exercise? I took a staggering number of comments and updated the BCP accordingly. Of course, some comments needed to be resolved, but the stakeholders were now interested, willing to meet, and genuinely interested in making the BCP the best it could be. Now that was a success!

What Worked

1. I took the initiative and drafted the best possible BCP using my knowledge of business continuity and what I had learned about the company.
2. Request a review and talk one-on-one with the stakeholders. In the scenario above, everyone knew those steps had taken place, so how could someone not know their "job" during an incident? Doing your due diligence and talking to all the stakeholders builds your credibility as a subject matter expert.
3. Bringing everyone together in the same room at the same time, physically and virtually, was an excellent opportunity to determine what was right and what needed to be modified in the plan.
4. Having subsequent tabletop exercises since, by nature, people forget what they've learned. In the first exercise, the object of this story was a huge spike in awareness about the BCP program and the fact that everyone in the room had a responsibility within the program. It is one thing to be aware of your responsibility but another to execute it. The ability to execute takes more time, which can be achieved through additional exercises. The diagram below shows the awareness and ability to perform (performance ability) I have witnessed for participants associated with business continuity programs after a series of BCP exercises. Notice that awareness and performance ability go down with time, so periodic exercises are needed to increase the team's ability to perform overall and maintain the level of performance obtained during the last exercise (Figure 5.1).

What Didn't Work

1. Don't expect people to read, ascertain, and comment independently on a comprehensive document without business continuity being part of their culture.
2. The company head is expected to possess the skills to orchestrate an incident with the comprehensive documents provided. They are used to having people do things for them, so while they want to be in charge and make decisions, they may not be the best person to coordinate the incident response. Therefore, a role was added to the plan as Incident Coordinator, to be appointed at the discretion of the IRT Leader.

⊕ Sharpshooter Tips

- Most people don't have original thoughts, and it is hard to envision something without something to compare or an example. Therefore,

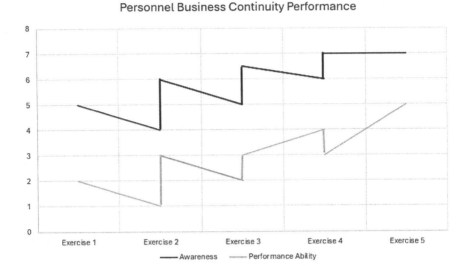

Figure 5.1 Personnel business continuity performance.

using the best information available, draft a BCP and let the first BCP exercise be a planned walkthrough.

• Have people in the exercise dedicated as note-takers. That way, you can focus on helping IRT walk through the exercise.

Misfire

• While this scenario turned out to be a success, consider having someone independently run your BCP exercise. Some companies specialize in disaster preparedness and do a great job conducting BCP exercises. I have found some exorbitantly expensive and others very reasonable, so shop around.

INSURANCE CONDITIONS AND CRITERIA

Resources you have are resources you should use

One primary way to address residual risk in an organization is to shift it to a third party, and one primary way that is done is through insurance products. Cyber risk is no different, and over the years, it has become not only more important to have a cybersecurity insurance plan in place but also increasingly difficult to procure at a reasonable expense. But it doesn't have to be.

Case in point, I had just started working at an organization that had never had to "do" much to get the coverage they needed. They essentially just let their broker know what types and amounts of coverage they wanted, and then they picked from a list of options provided by the various underwriters. But that was not to be the case this time around. The previous year, they had seen several major cybersecurity events in the industry and several large payouts from the insurance companies. The game's rules had shifted because of the impact of these events, and when the request was made, the response was silent. No one wanted to provide coverage for companies that couldn't offer specific assurances about their program. Unprepared for this contingency, the organization was backed into a corner and had to accept less than half the coverage needed at almost twice the previous year's price! The insurance gap required us to self-fund to meet contractual requirements and maintain certain levels of insurance with other organizations. Adding the additional premium expenses was a non-trivial amount of money that had to be set aside, causing a significant budget impact. Being blindsided like this should never have happened.

Sharpshooter perspectives

SSP1: I worked for a company that had a self-insured cyber policy. A customer required it. They paid a giant insurer $80,000 to put the insurance on the insurer's paper so they could give it to the client. The policy said my company was on the hook for up to $5 million. The issuer made a massive amount of money to do almost nothing! If you are a CISO looking for cyber insurance, be prepared to have another source of security requirements: your insurer. The insurers are reasonably good at quantifying controls that will reduce your company's risk of compromise. Essential tips:

1. Don't exaggerate your controls to the insurers.
2. Be prepared for them to utilize an outside scorecard service to track your security.
3. Use the insurer's due diligence list to help you create your security roadmap. The controls on the list will reduce the risk of compromise.

SSP2: Anytime I think of cyber insurance, I think of the Somali pirate attacks because when the insurance industry was creating the cybersecurity insurance market, the risk profile wasn't well understood, and they needed a model, so the risk of insuring ships passing off the coast of Somali was the closest model and the one used. It makes sense because there are strong parallels between a ship's ransom and a company's information. That aside, in my first role as CISO for a Fortune 500 company, the company's primary insurance carrier was the one that gave credibility to my risk program. The VP of IT did not accept my risk method, so with help from the VP of Risk, we scheduled a meeting with the insurance carrier. With the VP of IT present, the VP of Risk and the cyber insurance expert validated my whole risk program and said it was a model for all their clients to follow. So, as the first Sharpshooter noted, use the insurance company as a catalyst to improve your cybersecurity program.

SSP3: The small print taketh away what the large print giveth. Having wisdom and experience in the room during the contracting and onboarding stages is critical with any third party, particularly regarding DR, contingency planning, legal matters, and preparations for potential ransomware incidents and cybersecurity insurance. Significant trends have emerged globally over the past 5 years beyond any company's control. However, what we can control is our ability to establish a solid foundation for a proven, secure-by-design (cybersecurity framework) program. This includes implementing assurances, auditing, attestations, and demonstrations of hardening to demonstrate maturity. Such efforts can help negotiate lower rates, justify larger scopes of coverage, and address other variables related to third-party relationships. You cannot achieve this in a day, but often over an 18-month to 3-year period, so start planning and embrace your short-term needs to manage any immediate impacts beyond your control if you're behind the eight ball.

Target grouping

Insurance coverage is not a mystery, so don't make it one. There is no secret about what needs to be done to manage this essential risk mitigation component. Show up prepared, and you make everyone's job easier and more fruitful. In this case, we got what we prepared for, which was not what we needed.

In the following months, we discovered that many other organizations had experienced similar constraints in getting the required insurance coverage. I made it a point to reach out and have conversations with various underwriters to find out what they were looking for and where they saw the most significant potential risks. You would have thought I was showing them how to make wine out of water. They were shocked and thrilled to help and show me all the data they had to show where the most probable losses would occur and what could be done to minimize them. They also pointed out the many services in their programs that very little seemed to take advantage of. These include free tabletop exercises, framework reviews, industry threat intelligence, risk assessments, and more. Using those, I was able to grow capabilities within my program that were 100% aligned with the underwriter's criteria at a zero-dollar budget impact. It is a silver lining that didn't offset the poor renewal performance, but it is something to put in the win column.

I also got a virtual blueprint of how to structure a program that would be most appealing to an insurer. Armed with this information, we constructed a program "pitch deck" that showed how we covered the top issues critical for the insurers and highlighted the many additional aspects of the program that showed the holistic approach and forward strategic value of those efforts.

We were prepared and ready to go when we went for renewal a year later. The results speak for themselves. Our time with the insurance brokers was shorter, and the questions were almost non-existent. In the end, multiple insurers wanted our business compared to just one the year prior, and we ended up with well over twice the coverage needed at a price close to 50% less than the previous year's premium for insufficient coverage. It wasn't rocket science, and it wasn't a mystery. By involving our insurer as a partner and acting on the feedback, we locked in a solid win for the entire company that year. We continued to do so in the following years with only slight adjustments and updates to that original deck.

⊕ Sharpshooter Tips

- #1 NEVER lie on an insurance application. Omission, exaggeration, or otherwise.
- Create a "sales deck" of how you take cybersecurity seriously and what you are currently doing and plan to do in the future. You are selling your cyber risk program and using it to manage and improve it.
- Your insurance provider can be your best ally or another adversary. That is your choice.
  - ASK. Your provider will tell you what their most significant areas of concern are.
  - Take note and focus your program/efforts there.
  - A discount on cyber can mean a discount elsewhere.
  - Your risk profile is impacted across the portfolio, not just in cyber.
- The providers usually have free security resources that can enhance your program AND get you a discount on your premiums. Ask if these have proven effective in helping them avoid paying out on a loss.

⌐ Misfire

- Just because it worked in the past doesn't mean it will work today or in the future. Don't rest and rely on what you "think" should work, but instead, work to be ready so you don't have to "get ready" or be surprised by something you could have foreseen.

LEGAL PREPAREDNESS

₴ I need legal help now!

It was an ordinary Tuesday morning until the CEO received an alarming call from the CISO hired 2 weeks ago. The company had fallen prey to a massive security breach about a month ago. It had gone undetected until a large client reported that its dark web monitoring service had discovered their data with specific traces back to the company. The security breach exposed the personally identifiable information (PII) of nearly 100,000 individuals across all fifty states. Panic set in as the CEO realized they were woefully unprepared for such a catastrophic event from a legal standpoint. The compromised data included names, addresses, social security numbers, and financial information, setting off a legal and financial crisis. The company realized they didn't have the needed privacy and security privacy programs, and that is why they recently hired a CISO. Unfortunately, the company had not planned for such an event and did not have an incident response plan to address it. They knew state privacy laws were somewhat different, and now they had to develop quickly to comply with them. The General Counsel (GC) felt overwhelmed with the legal implications and didn't think he had the capacity or knowledge to address the matter quickly. Given the scale of the breach and the complexity of state-specific regulations, the leadership had to decide whether to rely on their internal legal team or seek external expertise.

Sharpshooter perspectives

SSP1: 100% go with outside counsel on this one. Even if the internal legal team is excellent, they are too close to the matter to give it the apathetic eye it needs. If the organization has cyber insurance, that would be the first place to look for recommendations around the IR and who third-party counsel might be. If the organization doesn't have cyber-specific insurance, still contact your insurance provider because of the existing relationship. Even if they don't cover a claim, they will have a good idea of who can provide the services needed and who they would have (and will) cover in the future. The best time to plant this tree was years ago. The next best time is today.

SSP2: That is a familiar story. The first place to start is if the firm has some cyber insurance. The insurance company usually has procedures to help work through the issues. Next, having had to work through something similar, you should go through both outside counsel and external incident responders. Usually, the large external teams have some notion of who can help in these situations. I was working for a similar large company, had an incident, and hired a well-known incident responder who helped handle the incident. After that, I worked with them to sign a zero-dollar incident retainer for the following incident. Unfortunately, the following incident came, and I discovered that the procurement department had left the incident retainer unsigned because it was a low priority compared to other work. I had to repeat the whole contracting process for the new incident! Follow-up to confirm that a contract is in place.

Target grouping

The CEO opted to hire an external legal firm specializing in PII breaches. This decision was driven by the need for expert knowledge of the varying state laws, the urgency of the situation, and the desire to demonstrate a proactive response to stakeholders.

The financial repercussions of the breach were staggering. The total cost included the legal fees and expenses associated with notifying affected individuals, providing credit monitoring services, and implementing enhanced security measures. The estimated final bill reached $15 million. The external legal firm assisted with filing and obtaining some insurance relief. However, there was still a significant dent in the company's financial health since insurance was capped at $5 million.

The external legal team navigated the maze of state regulations, ensuring compliance with each jurisdiction's unique requirements. They assisted in drafting and promptly sending notifications to affected individuals, minimizing the risk of further damage. Additionally, they liaised with regulatory authorities.

Based on lessons learned, the company invested heavily in the following after the breach:

- A robust cybersecurity strategy.
- Retainers with an external legal firm and managed security provider.
- A comprehensive incident response plan, including the nuances of different state regulations.
- Dedicated IRT.
- Cybersecurity training for all employees.

The company also fostered a culture of transparency and accountability, encouraging employees to report suspicious activity promptly. Regular drills and simulations were conducted to test the effectiveness of their incident response plan and ensure they were well prepared for any future compromises.

⊕ Sharpshooter Tip

- Don't underestimate the legal support required for a breach. Include your legal team in the incident response planning and testing. Consider having an ongoing relationship with a legal firm to assist should a significant breach occur. Like security and privacy staff, most companies cannot maintain the legal staff needed to respond.

⌐═╼ Misfire

- Being successful as a company and growing revenue quickly will attract the attention of those who profit from hacking. Governance must evolve with the business to manage the increasing risk at an acceptable level.

RECOMMENDATIONS FOR DIFFICULT SITUATIONS

⅀ Stand your ground

When leading the Computer and Information Security organization at a Department of Energy (DoE) facility, I was instructed by the Director of Security to provide non-approved personnel from a foreign state access to a location that held highly sensitive classified data. I responded, "Sir, this is not only a violation of law but also a violation of the Nuclear Non-Proliferation Treaty (NPT) overseen by the International Atomic Energy Agency (IAEA)." He responded, "Do it anyway." Faced with either breaking the law and international treaty or being insubordinate, I explained that I would not do it but would provide a two-page hierarchical summary of the legal controls for the area and why giving access would be a direct violation.

Within 2 hours, I provided the summary; he barely glanced at it and again instructed me to provide access. I refused and endured the threats and swearing. He said he was overriding me and demanded the procedure for providing access. I provided the procedure, and he stormed off. I documented the interaction, knowing that this matter was not resolved, and informed my direct boss when they were available the next day.

Sharpshooter perspectives

SSP1: One thing that I have found effective is a set of principles that lay a foundation for how information security is governed. "We comply with regulations" puts the foundation in place, and the other items are layered on top of it. The foundation, "We comply with regulations," I consider non-negotiable (Figure 5.2).

When starting a new role and asked how I manage security, I laid out this model, which allows me to call someone out later when they ask to do something outside of regulations.

SSP2: Classified data requires much more stringent controls and handling, especially in a system tied to critical infrastructure such as a government-based supervisory control and data acquisition (SCADA) system. Your response to

Figure 5.2 Foundational principles.

uphold the law revealed their non-compliance and desire to force you into an unlawful situation. I am concerned this is a verbal conversation. When legal matters arise at work, I, with discretion, always follow up sparingly on such situations with documentation over email with a structure such as:

1. Thank you for your time.
2. Here is what I understood and/or Action items.
3. Please correct me if I didn't understand this correctly.

Here is an example email:

Form 5.1: Example Email to Superior Verifying Concerning Direction

Subject: Follow-up on our Recent Conversation

Dear [Superior's Name],

I hope this email finds you well. I wanted to take a moment to request further clarification regarding our conversation earlier today to ensure I understand your instructions correctly.

During our verbal conversation earlier today, you asked me to [Describe the illegal activity your boss asked you to do], and you provided some context for why you believe this action is necessary. However, I want to confirm that my understanding is accurate and seek additional clarity on the following points:

1. [Specify the key points and details you discussed with your boss regarding the illegal activity.]
2. [Highlight any potential legal and ethical concerns that you raised during the conversation.]
3. [If there are any potential consequences or risks associated with the requested action, mention them.]

I would like to reiterate that I take our conversations seriously and want to ensure that I am following your instructions accurately. If I have misunderstood any aspect of our conversation or if you can provide further guidance or clarification on this matter, please let me know.

I appreciate your time and attention to this matter and look forward to your response. Please advise on the best course of action moving forward, as I want to ensure compliance with all applicable laws and regulations while carrying out my responsibilities.

Thank you for your understanding and support in this regard.

Sincerely,

[Your Name]

This email is an important way to document your understanding and a great way to shift liability to the other individual. They must take action to correct or document what they are saying verbally to avoid plausible deniability – especially for a supervisor asking you to perform unlawful acts. The problem is this action may raise their hackles and cause them to come after you. Going to HR may also do the same, so proceed cautiously, but do not break the law and hold fast to your convictions. When informing your direct boss, did you concisely record and document this in writing? Such sensitive matters professionally require sensitive handling and documentation, as described above.

SSP3: Was this considered a success? You didn't break the law and held your ground. It wasn't fun, but you navigated it and appeared to have the support of your supervisor. I would count that as a success in the face of an office bully attempting to cut corners by asking you to violate the law. How you define success in this scenario matters! How did your boss react to this, and did they coach you?

SSP4: The context of the DoE combined with "non-approved" personnel sent off immediate red flags for me as a specialist in counterintelligence, nation-state threats, and all-to-often insider threats of a contractor or individual that led to a breach.

SSP5: I hope your boss had your back on this. Having worked in the government, I have seen the same behavior, which makes me angry. I want to hear more of the story.

Target grouping

The Director of Security did provide access, and the foreign state accessed the sensitive area. Personnel action was taken against the Director, and he got early retirement. A few months later, I was promoted to headquarters. I consider the outcome of this situation a win, knowing that I was following my principle of always doing what is lawful, moral, and ethical.

Since all access is reviewed independently monthly, it wasn't long before it was questioned, and I was called upon to explain why access was provided. I took my documentation with me and explained everything that happened. While I was not directly involved, reports had to be written and issued at multiple levels within the government. During the whole investigation process, my boss had my back and provided valuable coaching. I also had support from the entire Computer and Information Security organization.

⊕ Sharpshooter Tips

- Have a predetermined set of principles and morals that you use to make decisions. These will determine the answer to difficult situations before you face them.
- Adopt a "Yes and" mindset instead of a "No or But." You can deliver the same information, but the outcome will be entirely different based on *how* you communicate it.

🔫 Misfires

- Being insubordinate even when you are right has the potential for negative consequences.
- Bullies will come after you seeking more power and can be embittered, insecure, and want to push you down. You'll need trusted colleagues, managers, and political power to navigate such threats.

CONTRACTS, SLAs, AND LEGAL ENABLEMENT

🔫 We signed what?!

Cybersecurity Manager: "Hey, boss. We've got a call from an angry vendor who says they won't support us because we blocked that remote control app."

CISO: "The unsupported, exploitable, non-monitored, remote control app that resembles malware?"

Cybersecurity Manager: "Yeah. That one."

CISO: "Tell them I don't care, and then tell them to use the enterprise-supported one that is fully compliant and secure."

Cybersecurity Manager: "I think you should take this one. They refuse to use any solution other than the old app, and their lawyer tells me we are in breach of contract."

CISO: {sigh}

I really, really wish that wasn't an actual conversation I had, but it was. After getting off the phone with a major supplier of support for critical systems in my organization, I had the team pull up the support contract in question, and sure as the sun rises, the organization had signed a contract 8 years prior that specifically called out a remote controlled application that had some severe security issues. In addition, the contract specified we had a duty to pay them not only the regular fees but additional damages should we intentionally obstruct and prevent them from providing said support. I was stunned and mortified at the same time. We were contractually obligated to either pay extra for a needed service that we wouldn't receive or have to open up a known and exploitable application to the internet and take our chances.

Sharpshooter perspectives

SSP1: This is a horror story waiting to happen. Surely, this contract isn't perpetually open-ended and must be renewed periodically and hopefully annually. Upon renewal, I'd force the use of a secure tool. If it is an extended time before the renewal, I would get GC involved to help explain to the vendor that while meeting contractual requirements is essential, it doesn't automatically absolve a third party from potential liability, including cybersecurity breaches. The vendor needs to formally understand that if they insist on a known unsecure solution to provide support, they will be held liable for any breach associated with the solution.

SSP2: Legal matters are often more about posturing than the law or what is right or just! Don't allow yourself to be bullied or threatened by another. Roosevelt said, "Speak softly and carry a big stick." This advice is the best recommendation I can give when dealing with such matters, embracing as well-rounded and experienced legal counsel as possible to understand all your options for the entire lifecycle of the immediate and long-term relationships and how to navigate the business relationship wisely. Ugly partners must be held accountable and will eventually reap what they sew. Your goal is to minimize the damage, and if you fail, fail quickly, move on, and learn from that mistake to avoid repeating it in the future before onboarding a vendor in the future of a similar nature.

SSP3: Given the organization's size, it sounds like a legal department should be on the company's side. I would start with them, outline the current contractual issues, and find out how much liability the vendor carries. I have been in this situation multiple times, and sometimes, it takes years to sort out until the contract expires. In the current common language environment for liability for security breaches, vendors find it harder to argue that controls should not be put into the contract.

Target grouping

This issue is, unfortunately, not uncommon. Many times, legacy contracts that have seen multiple renewal cycles often contain clauses or aspects that are either no longer applicable or downright wrong in the current environment. This case was one of those. This service provider was quite large and had an operational model requiring their employees to use their internal processes. This was linked to their employee training and business model – a rational enough reason but not something that made much sense. We were a "legacy" customer in their mind and had everything good and bad that came along with that.

After reviewing the contract and situation with the legal team in my organization, I got on the phone with our sales rep, their leadership, and their contract/legal team. They were 100% right in the black-and-white letter of the contract. We were indeed beholden to the specific technology listed there and the potential financial disincentives associated with interfering with that method. However, quite a few items were buried and missing in the same contract, which provided us with some leverage we could use. A lot can change in 8 years. Whereas we may have been "locked" into using a specific technology to ensure their ability to achieve the SLAs in the contract, we maintained control over where our data could be accessed and the accounting of the access that happened. The service organization had adjusted to an offshoring model outside the initially agreed scope a few years prior. At the time of the original contract, it was noted that none of the support team was outside of the United States, and that proved to be material enough change for us to have a more "productive" conversation with the vendor.

After explaining how they had missed producing adequate accounting of the access to our systems and the location of the resources performing that access, I noted that could not only be a breach of terms in the same contract they were trying to hold us to but had also kicked off an investigation into the third-party risk of continuing with them as a vendor after the current term that was set to expire in 12 months. Needless to say, a few more options became available, allowing us to start bridging the gap between us.

After several more meetings, we landed on a third option, which was different from our recommended solution. Still, one that had a sufficient level of security controls and auditability to satisfy the technical risk concerns my team had. This modern option was their current option with other organizations, but additional licensing costs were associated with its use and required updates to the systems being supported. The price difference was insignificant, so we agreed to pay the delta, and they decided to switch away from the old platform they had drawn the hard line on using. This third option became an alternate standard at our company because of the level of industry adoption, so it was a win–win in some sense.

I do have to point out that we got lucky a bit in this. The fact of the matter was that if we hadn't found weak points in the original language, it would have been a more painful and risky situation. It probably would have been a much more expensive solution as well. Because of this near miss, we prioritized creating an up-to-date Security Risk Addendum for all future contracts that would include the current requirements and clauses to ensure something like this didn't happen again. This addendum was added to every renewal negotiation and new vendor contracts, regardless of where they were instantiated in the organization. As renewals came up over the next few years, I was amazed and shocked at how many other places we were contractually obligated to allow risky situations. I frequently encountered new vendors who insisted that "no one else makes us sign this," necessitating numerous conversations to ensure we stayed on the right path.

⊕ Sharpshooter Tips

- The best time to have rational security controls is at the beginning of any third-party relationship. The second best time is at renewals. The third best is implementing them ad hoc whenever issues are discovered.
- Don't ever take a vendor's or partner's word for it when it can negatively impact your organization's risk posture. If something doesn't seem right, research and fight for the right thing. It isn't always easy, and sometimes it takes a little luck, but you are accountable for holding the line and fighting those fights.

- Consider annual capability assessments to validate that the third-party tools or organizations still provide the best value.

Misfire

- Never just "let it ride" when contract renewals come up. Change is the one constant in life, which is never in just one place. Make sure what you are signing today is relevant for today and not 8+ years ago. The issue should have been caught well before the time it did.

Chapter 6

Digital transformation

Cybersecurity must evolve alongside digital advancements. As technology grows more complex, so too should our defense mechanisms. The imperative is clear: adapt quickly or risk obsolescence. CISOs must anticipate emerging threats and adapt their strategies proactively. Success in cybersecurity transformation hinges on adopting new mindsets and tools and fostering a relentless pursuit of knowledge, all while maintaining foundational principles. It requires readiness to explore beyond the familiar and identify potential threats early. The effective CISO knows cybersecurity is about new technologies and nurturing mindsets to evolve with the changing digital landscape. This chapter explores the necessity of cybersecurity transformation in the face of advancing digital technologies. It delves into the importance of adapting defense strategies, fostering a culture of continuous learning, and the critical role of mindset in navigating the evolving cybersecurity landscape.

ZERO-TRUST ARCHITECTURE (ZTA)

🐍 Who do you trust?

I stared at the flickering monitor. Another data breach had occurred, and another hospital scrambled to contain the damage. Patient records were exposed, and lives were potentially at risk. It was a familiar scene, a recurring nightmare in healthcare cybersecurity.

The industry relied on a fortress mentality – a digital version of Fort Knox. Once a doctor, nurse, admin, or even a volunteer was "in" the system, access was more or less wide open. It was a system built on trust, a trust that had been repeatedly shattered.

I quietly shut my laptop. Our organization wasn't that much better, and this model wasn't working. Who should have access to a patient's X-rays? An oncologist, yes. A billing clerk, definitely not. However, the current system granted access based on roles and broad strokes that painted a blurry picture. We needed something granular that mirrored the ever-changing needs of a hospital environment. "Identity has always been the security perimeter," I repeated. It wasn't about walls but who got the key and for how long. The default trust of legacy systems wasn't working.

As I pondered this, my phone buzzed, and the message indicator light blinked. Getting a text is not uncommon, but the time of night and the coincidence of timing gave it more urgency than I would typically have assigned it. The message read, "We have live connections with Company X that were just breached, and the service account assigned to them is actively being used."

Sharpshooter perspectives (SSP)

SSP1: Connections to third parties have always been one of my most significant concerns because, essentially, my network and systems accept the security posture of the third party over which I have little to no control. If you are forced into a third-party connection, here are some must-have controls under the access restriction category.

1. Limit access to only what is necessary. For example, if the third party only needs access to one application, lock them down to that application.
2. Enforce named accounts and use strong authentication methods such as MFA.
3. If access is only needed occasionally, keep access turned off until needed.
4. Frequently audit to ensure that access permissions are still needed and valid and that accounts are used only for their intended purpose.

SSP2: Securing medical environments is inherently complex due to several factors. These include the involvement of numerous stakeholders with diverse requirements and needs, a culture where security has often been overlooked, and the presence of a massive amount of Internet of Things (IoT) devices. Many of these devices have little to no security or are grossly insecure, making them easy targets for exploitation. This results in an enormous attack surface for the entire network. A significant challenge we face in 2024 is securing the most complex cyber assets in human history. This includes legacy computing, new cloud computing, and intricate hybrid infrastructures scaled across global networks. This task is unprecedented in complexity, often involving dozens of providers, thousands of applications, and countless IoT devices, with innumerable objects, people, and data constantly moving. Are you overwhelmed yet trying to secure all of this? Zero-Trust Architecture (ZTA) and embracing newer technologies and transformative solutions like software-defined networking (SDN) enable segmentation, isolation, and cloud security controls. When properly configured and managed, these technologies provide excellent granular security controls beyond the legacy controls in traditional computing environments. When ZTA is appropriately designed and supplemented with advanced security controls, complex environments like those in the medical field can achieve identity access management, data security controls, and other requirements necessary for adequate security. This security level is impossible with traditional tools, techniques, and procedures. If you haven't transitioned from legacy computing yet, it is imperative to do so, as the evolving technology, architecture, and global realities of 2024 necessitate this transformation.

SSP3: Every environment I have ever been in had access control issues. The first access management project I was involved in aimed to "provide employees with the right access at the right time." Most access control projects are not implemented because of the cost of hiring personnel to maintain the systems and processes. We have solutions for access control, including vendor access, and the gap we always face is for people to manage the systems. The first step is to understand who governs access control. If governance is within the IT operations or service desk teams, the strategy will be to provide access to reduce calls, not security risk. This strategy leads to over-permission. The next piece to understand is whether the executives support an access control project. If management always bows to doctors or other employees, sadly, it will probably take an incident to gain support for an access control project. Hopefully, both areas will be covered in this story so that suitable controls can be implemented.

Target grouping

Legacy systems typically rely on macro controls, like having a single security guard oversee an entire hospital. In contrast, ZTA employs micro controls, effectively placing a security guard at the door of every room. This approach creates a finer, more adaptable security mesh that can adjust to an organization's changing needs.

I knew implementing even a portion of it wouldn't be easy. There would be pushback, a reluctance to let go of the familiar, even if it failed. But we needed a model that operated on the principle of "least privilege." Where every access request was a rigorous interrogation, each login was a fresh check-in. There are no more permanent keys, only temporary permissions for specific tasks. Not just who you are (doctor, nurse), but what you need (specific patient file) and why you need it (treatment plan).

In the months leading up to this specific event, I had been working out a plan to get our organization from where we were to where we needed to be from a security perimeter and controls standpoint. Like most industries, modern healthcare has seen the security perimeter change dramatically from a specific place to just about every place. The need to access resources and the ability to support the delivery of value and care, no matter where that happens to be, has been moving rapidly to table stakes. Bringing the perimeter down to an atomic level, where each entity needing access was isolated and controls enforced at the level of the specific resource needed. Nothing else was going to be critical to doing that. ZTA could make the model of that "default deny with access only granted by exception" a reality.

With that in mind, one of the only ways to get the plan I had put in place was to ensure the organizational and operational impacts were dramatically minimized. Change is hard enough for individuals and organizations, so implementing what amounted to a complete culture change (from granting access to everything and denying by exception) and technology adjustments must be treated delicately. Service accounts and vendors represented one of the most significant risk vectors and the entities we had the most control of and could impact immediately. So that is where we had been working, and it proved to be our saving grace in this instance.

Before implementing our ZTA strategy, service accounts were often reused for multiple services. For vendors, that sometimes meant they would get access to a service account to do one action or capability, and there was often a dramatic expansion of access scope because that specific account got used elsewhere. We started with access via VPNs and slowly ratcheted them all down to minimum needs and documented the associated accounts and systems. We then spent time creating specific service accounts for the services and locked them down to the least privileged across the board. This caused a dramatic increase in service accounts, but it also caused a significant decrease in associated risk.

So when I opened my laptop back up and dropped into the chat channel with the incident team, I saw they were already on it. They had identified and disabled the specific suspect service account and were validating the traffic on the VPN before moving it into a disabled status. They were able to confirm that the service account was only associated with a minimal number of systems and that disabling it would have no impact other than on the specific use cases related to Company X. The VPN had been trimmed down to only allow the minimal ports and any other traffic had been stopped at the door. I thanked the team and observed the shutdown of the VPN. Afterward, when reviewing my notes, I noticed that this specific account and company was one of the overly permissioned ones that had seen a massive decrease in access only about a month prior. If we hadn't already been on the ZTA path, I am unsure what would have happened.

⊕ Sharpshooter Tips

- Identity is the security perimeter, which should only be opened by exception on a case-by-case basis and closed as soon as it is no longer needed.
- Start with the entities and identities you have the most access to and can control. Vendors and other third parties should be priority targets.
- Do not try to do everything at once. That is a recipe for failure when implementing ZTA. Current business processes are based on current

infrastructure and security models. Change them too quickly, and your efforts will quickly be shut down.

- ZTA will one day be the self-evident standard, but until then, it will feel like you are swimming up a waterfall and trying to explain it to folks.

⌐⌐⁊ Misfires

- Never reuse service accounts. One service = one service account.
- Do not over-privilege service accounts. Unfortunately, this is common because giving them more access is often much easier. After all, identifying all the interactions they might need to perform at a system level takes a lot of work.
- Do the work and figure it out. The future you will thank you.

CLOUD FIRST ADOPTION AND MIGRATION

⊇ Why are we in the data center business?

The tense silence of the Network Operations Center was like a heavy blanket had been dropped on the entire room. Panic flickered across the faces of the engineers as the primary data center status board blinked an evil red. Minutes became an eternity as they desperately tried to reach the secondary facility. Then reality started to sink in: the primary data center was hard down, and the secondary was unreachable. Both data centers, the supposed failsafe fortresses of our digital lifeblood, were down. An unknown facilities issue had turned our entire IT infrastructure into a cold, silent tomb.

As the clock continued to click, it became frighteningly clear that this wouldn't be an easy or quick fix. Amidst the chaos of recovery and business continuity that stretched from hours to days, many things came to light. People, processes, and technology were a mixed bag of good and bad. However, one thing was clear: few elements could be considered outstanding or modern in our data centers. We were a well-oiled machine, but that machine was designed for a bygone era. We weren't in the data center business; we were in the healing business.

Historically, legacy institutions are a fortress of data. Financial information, patient records, personal information, you name it – everything residing within the confines of their own data centers, a physical manifestation of control. But the world has moved on, and so have the threats. Solutions are no longer a single, monolithic instance of technology and data that can be locked away in a room. Instead, the modern world is increasingly built on composable services that require hybrid connectivity to implement, operate, and support at a speed and scale. A speed and scale that requires focused talent and resources to maintain. As a security leader, I knew clinging to outdated practices was a recipe for disaster and could not provide the level of resilience needed going forward. The irony hung heavy in the air – why were we still in the data center business? Why waste resources replicating what cloud providers could do better, faster, and frankly, more securely?

Sharpshooter perspectives

SSP1: We all fear this moment when we don't realize our accepted risk. I don't know the whole story, but I know a few things that would have averted it! Demonstrate proactively that you have disaster recovery (DR) planning and that it works. Are you testing recovery and backup every month? What third-party relationships do you have? How are those structured? What dependencies exist? How have you accounted for those in your DR planning and preparedness to ensure you have what you need in your shared responsibility? It sounds like you shifted responsibility too much and figured you were covered with your cloud and network providers when, in fact, you were not. It was not stress-tested ahead of time for this type of disaster. It is better to plan and prepare in advance and learn from the lessons of others than to ignore, neglect to investigate, test, evaluate, apply due diligence, or make assumptions. Otherwise, you may only discover in a crisis that you're significantly at risk. Cover your A**ets.

SSP2: I want to understand how the situation happened in this story. Most upper executives have heavily pushed for cloud strategy for every company I have worked for except for the defense contractors. Usually, it does not save money, but it adds several capabilities that aid recovery and availability.

SSP3: We are still in the data center business because it is cheaper. I run a hybrid environment and use a decision tree to determine if something will be hosted within my co-location data centers or on a Platform as a Service (PaaS). It would be nice to go completely cloud, but there is a very dominant limb on the decision tree called "cost." The cloud was initially advertised as very cheap, and in some cases, it is, but many solutions would cost up to 50% more if hosted in a cloud.

Target grouping

In the past, the idea of having a corporate data center where all IT resources were located, supported, and managed was the norm (and in some critical sectors, that will probably always be the case). The costs to stand them up were enormous, though, and their operational aspects required skill sets and talent profiles that were often wildly divergent from the business they were put in place to support. Fast forward a few decades, and running your own data center, your own private "cloud" where you keep everything, is becoming far less in fashion than renting space in someone else's. Cloud computing is just renting time on someone else's resources in someone else's data center. An infrastructure that they are 100% dedicated to running and operating at peak performance as their primary business. Organizations not focused on the specific needs of running a data center struggle to manage every aspect at a high level, especially when those resources could be better utilized elsewhere. Each year, it makes less sense to maintain control over all data center operations unless specific compliance or safety reasons, such as in the military, require 100% control. Otherwise, the effort and resources invested will likely yield less value than outsourcing, pulling you away from your organization's fundamental mission.

To stay relevant and growing, organizations need to develop a strategy to determine the most appropriate place to put resources. Unless you are in the data center business, running your own data center at a large scale is not the best place for your resources. I am not suggesting that all functions move to the cloud or that we should eliminate private data centers. We should first look at modern architectures and hybrid options and balance on-premise and off-premise resources to concentrate on core business functions. At the very least, organizations should look at co-locating facilities that handle everything that can't be a core competency. That co-location may be putting physical servers at a location purpose-built and maintained for IT systems or using a private or public instance at an established cloud service provider. Keep what needs to be on-site local, but everything else should be placed where it can be best maintained and supported.

So, what does this have to do with being a CISO? Well, protecting the data and resources of your organization needs to happen wherever it is being transacted, transmitted, or stored. That location is probably better someplace other than your walled garden of slowly decaying infrastructure. The cloud isn't some temporary space. It can be a secure, scalable network constantly updated by security professionals and not overworked IT staff. The organization wouldn't be managing physical machines but configurations, processes, and a modern infrastructure described by code, not legacy metal. Identity, not a physical perimeter, would be our security model. Zero Trust, a concept where constant verification trumps blind trust, would be our watchword.

It took a lot of time to get everything back up and operational. We scrambled and enacted our DR plan, but the truth was that our on-premise data center was a creaking monolith, held together by duct tape and overworked engineers. The cost to recover was staggering. Many systems didn't come back up, and restoring operations to total capacity was more about replacing and updating infrastructure components at the data center than ensuring databases and solutions were working.

In the aftermath, the board demanded answers. They were tired of the constant disruptions and the ever-escalating IT budget. There were many contributing factors, but the actual cause of the failure was a physical component at the facility that had deferred maintenance. Something that would most likely not happen at an organization whose primary business was owning and operating data centers.

This event was enough to cause a shift in the organization—to a cloud-first approach. Our on-premise network, once a physical barrier, became a bridge to the cloud. We embraced Software Defined Networking (SDN and PaaS solutions, creating a hybrid architecture. Our physical data center remained, but it was moved to a co-location facility that specialized in that work and it started to look like just one node in a larger, interconnected landscape. The logical network, encompassing cloud and on-premise, became our new frontier.

Cloud adoption wasn't without its pitfalls. Agility, a significant selling point, could be a double-edged sword. Spinning new cloud instances was fast but could also lead to security shortcuts and budgetary impacts. We had to balance leveraging the cloud's speed while adhering to strict security protocols. Security became much less an afterthought and much more the foundation upon which our cloud journey was built. But that is a story for another time.

⊕ Sharpshooter Tips

- You can do everything right from a technology and controls standpoint but still struggle with business continuity if things outside your control fail.
- Stay modern in your approach to the organization's security and infrastructure, which you rely on to operate your role's capabilities and functions.
- Sometimes, it takes a major event to move people from the old way of doing things to a better way. Have an idea of what that new way looks like so you're ready to guide the conversation when the opportunity presents itself.
- Monolithic architectures should be avoided unless required. Modern architectures are composable solutions and are often much more resilient and agile.

⌐‿ Misfires

- Stick to your strengths and specialties. This counts at an organizational level as well.
  - Don't pretend to be an expert in something if you can't be that expert and your business relies on that expertise.
- Don't underestimate the physical risks to your technical world.
- Small things can have significant impacts, and deferring the "non-priority" items for too long can create high-priority problems.

⊇ It isn't work from home. It's work from anywhere

The office was just about empty, with the only noticeable sounds coming from the HVAC systems circulating air at a constant and comfortable temperature like a white noise metronome. A single employee sat at his desk while I typed away in my office. Emails and virtual meetings were the flavor of the day as I filled the hours of this otherwise average weekday. The previous week, I sent all the employees home to work remotely for a disaster recovery and distributed work experiment. It wasn't a response to any mandate or an effort to recruit and retain talent. It tested whether the team could operate at expected levels while geographically distributed and remote in the case of a natural disaster that prevented them from coming into the office. The employee in the office with me had asked for an exception to go to the office because he felt too distracted at home and couldn't accomplish the tasks that he needed to do. I felt a need to be still on-site and "lead from the front," leaving the two of us as the only residents of the entire floor while the business continuity exercise played out.

Things had started a bit awkward, with the first week being just this side of chaos, but by day 7 of the 14-day experiment, an interesting trend began to appear. Operational effectiveness increased while ad hoc conversations and collaboration took a nosedive. People started to communicate at all hours of the day, and "work hours" blurred into a blend of all days and differing circadian rhythms. Still, work satisfaction metrics jumped higher than I had seen in over 20 years as an executive leader.

Little did I know that less than 12 months later, the entire country would be locked down and scrambling to figure out how to keep business flowing while being anywhere but the office.

Sharpshooter perspectives

SSP1: You'd make a great captain who goes down with his ship. Did you go down with your network during Covid, or did your network survive thanks to some proactive preparation? As I read the story, it's clear that the CISO is leading by example and fostering a community of loyalty with practical outcomes. Staff often become "busy" and accomplish nothing that moves the needle in the workplace, directly reflecting poor leadership. If you have busy people who aren't effective, look at yourself and your culture and figure out where you can impact it to move the needle. Excellent work ethics, commitment, and loyalty start with trust and relationships. What are you doing as a leader to connect with your people, professionally and personally, and take one for the team and be that leader in the office, anchoring the team so others can be home with family on holiday? Lead by example, foster mutual love and respect, and see team spirit, commitment, and retention rise in the workplace.

When an organization takes the time to perform disaster preparedness, run/playbook incident response, and impact studies with the right stakeholders for the right threats, the most important outcome is what the organization does after the exercise. The exercise raises awareness and collaboration but doesn't address process improvement, implementation of new tactics, improvements in policy and procedure, or proactive hardening and preparation for a disaster. Due to politics and a lack of leadership, many organizations "get busy" and do little to nothing following such exercises, failing to lead and proactively reduce risk. Great leaders are great planners! Immediately following exercises, set expectations for an after-action report, recommended actions to be reviewed by key decision makers, and a road map for short- and long-term goals to improve operations to reduce risk. "If you fail to plan, you are planning to fail." – Benjamin Franklin.

SSP2: What a great leader to test the ability to continue operations with a distributed workforce. The great social experiment in 2020 showed that many people can be highly effective while working remotely, and some cannot. I applaud the professionalism of the worker who self-identified that working from home was not a good option for them. When it comes to IT workers, management expects that they can work from anywhere. Let a critical system go down, and you'll see people working extremely hard from home to get the system back up! So if they have to work from anywhere, why can't they work from anywhere? I am amused by the old-school CEOs who require everyone to return to the office since "we cannot be productive when remote," and they don't recognize that everyone is remote when you have people working together but distributed in offices around the globe. They have work processes and practices that work for multiple offices, so why can't it work with practically everywhere being an office? These same CEOs said, "It will never work." IT leaders showed that it could work. The narcissistic tendencies of CEOs needing to think they are in control drive these decisions and not meaningful data. They love to quote the studies that support their position, even when they aren't relevant while ignoring numerous studies. Quite frankly, I have been a remote worker since leaving a high-security job almost 20 years ago, where everyone had to be on-site working shoulder to shoulder due to the security requirements.

SSP3: I have had to fight for people's ability to work remotely several times, long before Covid. I am ecstatic that you did this as a business continuity exercise. Senior management frequently asked me, "How do you know what they are doing?" In the job description, I would explain that I measure my work by product and duties. From some execs, this statement perplexed them even more. If I have someone whose duty is to review vendors' security, I will notice that the reviews are not getting done. I also review work products (vendor reviews) with the employee. This usually leads to another question: "How do you know if they are busy enough?" The answer is a couple of ways: (1) I usually do most of the duties each employee has at some time and have an idea of how much work each entails; (2) I ask the employee how things are going. In addition, I usually have a set of metrics that I am monitoring. The thing about most security professionals is that they are curious and are typically concerned about a broad number of risks. Most of the time, I don't have issues getting them to work; the real issue is getting them to work less because burnout is a real issue in the profession!

Target grouping

If the pandemic taught us anything about work, the traditional ideology that all roles required you to be at some centralized location to produce valuable work products was untrue. In many cases, the ability to effectively execute a function must be enabled from anywhere (geo-location independent) to generate value. It is one of the core concepts behind ZTA, where the location of the entity doing the work doesn't matter. Still, the identity of the entity and the ability to enable the work to be done regardless of the geo-location is what matters.

Information security is one of those disciplines that falls squarely into the realm of functions that do not require a specific geo-location for the work to happen. The ability to acquire and utilize resources dramatically increases when you remove the requirement for constant geo-locked talent. Yes, it is still better for collaboration and knowledge transfer for people to be in the same physical space, but it is not something that needs to happen all of the time. Suppose the focus is on enabling the work to occur regardless of where or when it is done. In that case, the resources tasked with doing it can be force multiplied and aligned to increase effectiveness and worker satisfaction. Then, when you do need to have your team work in a geo-specific location (for team building, collaboration, and alignment items, as an example), the same level of capability can be accessed and enhanced to create a culture that focuses on the result, instead of the legacy mechanisms on how to achieve it.

When the information security function is structured this way, enabling other business areas to do the same is much easier. To do this in cybersecurity, you need secure, compliant, and privacy-enabled systems and processes in place. However, when those are ready, the Security by Design principles and models can be adapted to other areas where this could be an option. It is a "win–win" situation for the larger organization and the employee.

The nature of business and work is different from what it used to be. The ubiquitous nature of connectivity and the inherent composability of modern solutions that take advantage of that have established a new norm for both business and cybersecurity needs. Any CISO who hopes to have long-term success needs to embrace this and structure their programs around it, or risk becoming ineffective and ultimately irrelevant to the needs of modern organizations and work environments.

Cybersecurity needs are everywhere, all at once and all the time. Shouldn't your program be structured to address those needs from both the systems and the workforce?

⊕ Sharpshooter Tip

- Legacy models and concepts must be reviewed and reevaluated regularly. Work location is one critical area. Enabling the workforce to "work" from anywhere allows the ability to recruit and retain talent from a larger resource pool. It also builds a higher level of resiliency and adaptability for the inevitable emergencies that could make a location-dependent function unavailable.

⌐⮞ Misfire

- Not everything legacy is terrible, and not everything new is good. One of the unintended consequences of geo-diverse working is a decrease in team connectedness. Having a plan or model in place to address the personal dissociation that can happen is critical to the long-term success of your program and team cohesion.

⧉ Iteration is innovation

"Where is the 'Wow' in the Board deck? I'm having trouble finding it." I stood there, a little stunned by the question. We had taken several weeks to work on a concise, understandable, and informative set of slides for the annual planning meeting with the Board of Directors. The cybersecurity program status was clear, and the projected goals for the next 12 months illustrated a rational and risk-reducing plan that would increase organizational compliance without any significant increases in budget or staff needs, but for some reason, all that was being overlooked by the senior executive who I reported to at the time. Knowing I'd probably be sorry for asking, I decided to go for it and say, "What do you mean by 'Wow'? Looking at this, I see quite a bit to be excited about." A raised eyebrow and a more stern than it should have been voice replied, "When we hired you, we were looking for someone who could help kick our program into high gear. I don't see anything groundbreaking here that would lead me to believe that. What secret sauce will we show and sell to the Board to get them to increase support this year? What solutions must we implement to stay ahead of the bad guys and the competition?" I knew where this was going, so I took a moment to gather my wits about me, then said, "The secret isn't in the ingredients. Like any good chef, the innovation is in how you use the ingredients you have to work with." Intrigued but still skeptical, he said, "Alright. Show me."

Sharpshooter perspectives (SSPs)

SSP1: All too often, leaders are myopic, focused on a specific short-term metric, goal, or achievement instead of a long-term transformative process. CISOs win the cyberwar one battle at a time, with a war plan in place, always keeping perspective and the battlefield in mind. Great CISOs have a comprehensive situation report (SITREP) integrated into their operations, KPIs, and metrics. They structure project management into bite-size tasks that ensure success and enable the team to operate at the speed of business with clear priorities. Achieving this at scale in a high-paced, complex technical environment is much harder than one might think, yet it is a challenge every CISO must now embrace! A risk-based approach is the core foundation for all prioritization, ensuring the best outcomes in long-term success. For example, current conditions in patch management exist where bad actors, on average, have exploited new patches within 14 days while companies strive to obtain "N" state patching over a 30-day day. You are at risk with every applicable patch, and your highest-value assets hang in the wind 50% of the time. What are you doing that is risk-based to ensure your most important assets open to exploitation are patched first, in days or weeks instead of months? Embrace the principles and practicality of continual learning and improvement towards the essentials of risk management to proactively reduce risk and recover quickly when you have an incident.

SSP2: This makes me think about how I approach potential security products. Here is my thought process.

1. Does it address a significant risk in my risk register?
2. Did I miss a significant risk if it isn't on my risk register?
3. Does a solution I own but haven't implemented solve the same issue?
4. If this vendor will solve one of my significant risks, is this the right vendor? Are they hoping to make a big splash and get acquired by one of the big security dogs? Does my contract protect me during a buy-out?
5. If I buy the solution, do I have the staff required to implement, maintain, and use it, or will it become an expensive boat anchor?

By considering the implications of these questions, one can sort through the vendor hype and help avoid a potentially costly purchase and diversion of time to nonproductive activities.

SSP3: My most significant "Wow" moment was when one of my security engineers installed our first FireEye appliance in 2011. We had a bet on how many infected devices we estimated we had in the environment. He bet 10. I bet close to 100. We were both wrong, and I bought the engineer lunch for being the closest. The first day the application was plugged in, there were no alerts. On the second day, there were no alerts. Finally, on the third day, one of our employees hit a drive-by-download, infecting their machine. We were lucky to have no infected devices because absorbing the new technology took nearly a year. Based on the intelligence from the FireEye devices, we restructured our patching to patch frequently attacked software. We restructured our URL filtering firewalls to catch more. We also reworked our incident response program. Overall, one "Wow" technology took us at least 2 years to integrate fully. The biggest lesson from all of this was that our current patching program caught a lot of attacks, which appeared to be why we had next to no infected devices in the network. There is no "Wow" in patching, but it proved to be the most critical control in our arsenal.

Target grouping

While "showing them" the "Wow," I covered four primary points to drive this home:

1. **A Secure Future is Fueled Through Daily Iteration**

   As CISOs, the allure of rapid innovation through sudden technological advances constantly tugs at our attention. The industry's breakneck pace seeds the illusion that security strategies must match that volatility by reinventing continually. However, the most enduring innovations strengthening enterprise cybersecurity take shape through systematically iterating existing approaches, not chasing sparks that flash and disappear.

   Look at multifactor authentication (MFA), for example. Now a security staple, it arose from decades of incremental enhancements to access controls – password complexity policies giving way to one-time pins before the eventual effectiveness of biometric factors compounded MFA into the gold standard it now represents.

   The roots of any seismic shift generally trace back to iterative progression built up behind the scenes until a tipping point into mass adoption triggers. In cybersecurity, this means recognizing and doubling down on the power of recurring marginal gains made over long timeframes.

2. **Compounding Gains are Achieved by Tuning Out the Hype**

   Over my tenure as a CISO, the single most dangerous distraction gnawing at essential progress has been hype over the latest bleeding-edge security solutions. It's easy to get swept up in the constant barrage of vendors pledging revolutionary protection powered by AI or quantum-safe algorithms.

   However, authentic innovation in security arises more from pressure testing existing controls and processes rather than rip-and-replace implementations built on promises that are not proof. Actual transformation compounds not from one-off deployments when threats emerge but daily stress testing of the fundamentals currently in place.

   Staying focused on those less glamorous but crucial iterative security layers bears real fruit if it persists steadily over the years. Sort of like the "cornflake effect" of Yankee Stadium being built by transporting and placing small loads of debris day after day until the massive stadium took shape through sheer cumulative effort.

   Similarly, something as basic as mandatory password changes done quarterly since the early 2000s led to significantly more hardened authentication landscapes now made possible by stronger, secondary controls branching from that initial seed.

3. **Directing Iteration Through a Risk-Based Lens**

   Managing security through an iterative rather than reactive lens requires constantly filtering efforts through a risk matrix focused on likelihood plus impact. This method allows for accurate targeting of controls and processes that need priority attention because risk reduction potential is highest there.

For example, recurring penetration testing around high-value data assets prioritizes one-off pen tests on less sensitive systems. The volume and velocity of testing on crucial areas reveal more vulnerabilities over multiyear spans, allowing much more significant risk reduction through iterative remediation.

That risk-based approach also includes consistently reevaluating assumptions on existing controls as part of critical iterative cycles. Regularly question the status quo on current processes with the same zeal as vetting new technologies.

Tuning out hype always starts by cultivating a learning mindset on previous decisions, not a rigid belief that those choices remain permanently sound.

4. **Confronting Complexity through Consistent Steps**

CISOs need to pay special attention to recognizing innovation, not in implementing silver bullet offerings but in doubling down on amplifying risk reduction through recurring, incremental improvements, particularly focused on known critical assets.

Embarking on massive digital or cloud transitions may seem mandatory to rapidly advance defenses, but teams left bruised by unsuccessful big bangs often regret skipping the less flashy block-and-tackling required to shore up controls on legacy layers before trying to re-architect them.

True innovation arises from a fusion of knowing precisely which small iterative actions today manifest into maximum impact down the road while accepting that the exact shape of what emerges may shift as progress unfolds. But staying doggedly devoted to those marginal gains tees up seismic change in time, even if initially confined to existing technical debt.

After all, iteration is born from intent and resilience, and the engine generates lasting innovation. And that's what will continuously harden the next generation of cybersecurity. Unfortunately, after all that, my boss at the time was more interested in the flash of the solution than the core of it and decided to purchase a "cutting-edge solution to accelerate the environment to the next level rapidly." Not sure if they needed to spend the budget or if something else was at play, but 2x the money went towards adopting a new solution that turned out to be a massive failure and ended with the executive "finding a new opportunity." You can't win them all, I guess.

⊕ **Sharpshooter Tip**

- You can do everything "right" and still not get the desired result you are hoping for.

⌐╼╤ Misfire

- The more you know the hidden and nonobvious motivations of the stakeholders involved in your program, the better off you will be. By not recognizing that the criteria for success were significantly different, achieving the hoped-for success in this instance was hard. Do the pre-work even if you think it isn't needed.

⊇ Agile development shortsightedness

As a newer leadership member of the team, I focused on listening and learning in the first 30–60 days while building trust. I was especially focused on the data in the SIEM, as I needed to process the information in both structured and unstructured manners to achieve consistent, automated, simple queries such as IP and WHOIS lookups, reputational queries, correlations, enrichment queries, and malware attributions with reasonable levels of trust.

I observed a wide diversity of tools and tactics in SOC operations and a need for more consistency and understanding amongst all staff, seeking to standardize the core foundation of SIEM processing to help the SOC focus on analysis instead of click-fatigue lookups and queries. The leaders of SOC operations were "chief doers" who earned their role by working harder than the others, not necessarily individuals with people skills or leadership cultivation and mentoring.

The Agile Development Team, who had largely been with the company for over 10 years, were still young, worked odd evening hours, and often from home. They informed me that we had state-of-the-art artificial intelligence within our tech stack. Documents of all kinds were dropped into an unstructured database. They could easily be searched and retrieved without any effort, quickly and on a massive scale, according to the Agile team. Agile team members were motivated with monthly management by objective (MBO) metrics for bonuses, focused only upon short-term requirements.

Sharpshooter perspectives

SSP1: Listening and building understanding is a great strategy when establishing yourself in any new security leadership role. Since a firehose awaits one in any new security role, finding the right balance between listening and taking action is a difficult judgment call that is enhanced with experience.

Playbooks with defined process flows are essential in any SOC operation. Create the list of playbooks needed and focus on developing and implementing the most critical first. In critical incident situations, a database of documents that can be searched and enhanced with AI capability will never replace the ability of a seasoned professional who can instantaneously decipher a problem and initiate the proper containment.

Knowing the author is a seasoned incident response and remediation professional, I predict that after assessing the tools and ensuring they are adequate and that playbooks are in place, the focus will be training and skill development for the SOC staff, followed by team building.

SSP2: Cultivate a culture of "automation first" in the SOC team "doers" so that they can be trained to manage, mentor, and train staff rather than spend time clicking through incidents. Based on the use cases provided by the SOC team, the Agile Development Team should be able to automate specific process flows in the SOC. If the team has issues developing this, they may not have the right technology or skills.

SSP3: Huge maturity red flags here. Competent people engaged in heroics (ad hoc) is what this sounds like to me, combined with a total mismatch of incentives to actual work needed to address the issues. The culture issue is one of the first things that jumps out at me. In immature programs, it is not uncommon for the most technical to get the lead role and be escalated into management. It is also one of the biggest reasons these managers struggle with their roles. Management and technical expertise are wholly different, and different talent profiles are needed to succeed. This is one of those situations where everyone needs to come to the table and have some hard conversations. What works in a "Main Street" organization does not scale to a "Wall Street" level without significant standardization, role clarity, and leaving the urge to be everything to everyone behind. When organizations grow too fast without considering factoring solutions into repeatable patterns, they are bound for issues and avoidable chaos.

Target grouping

In hindsight, I was hired to acquire a company to buy versus build, and the purchase fell through. My role was never to fix and develop the solution in-house. I was left with broken pottery, and management wasn't all-in on helping me fix the problem. Not one to fail, I quickly recognized the actual root causes, making it impossible to succeed: cultural challenges, lack of leadership, and transformative gaps.

You will fail without a vision, actionable plans, and a proper foundation. A roadmap was created for the prioritized data ingestion, parsed management, storage requirements, and how information can be processed intelligently for the intelligence and MSSP operations lifecycle. Once this plan was in place with all considerations, we bit off what we knew we could, slowly untying the Gordian knot that was in place in operations that was created over 10 years prior.

We started with 93% of data being blobbed, executives being snowballed, and smoke and mirrors from the DevOps team. After a 3-year transformative journey, we ended with 100% of all data being parsed, then enriched and correlated in an SOAR, then automated through multiple tools and lookups formerly performed by staff, saving thousands of dollars in monthly productivity and consistency in ops. We also ended with truth and transparency instead of what we started with, a true transformation, reflecting a cultural change where we learned to accept our current state and work towards maturity, celebrating our current state and maturity gains instead of trying to make everything look good despite the current state – no matter what that state is.

Automation was one of the most challenging processes due to the complexities one learns when attempting to automate. You might think you can easily automate things until you try, and then you realize the process is not documented, and there are twists and turns you didn't expect – resulting in a mess. In short, automation is challenging to turn out consistently!

Start small with automation, streamline processes, tooling, people, and technology. Audit it heavily until you have the expected results, then scale up. This is critical as you scale projects with additional tooling over an extended period. Embrace the crawl-walk-run concept so you don't get overwhelmed with all the problems that automation can reap, focusing on the more accessible automation elements within your project first. In this case, what started with simple query lookups for IP and domain, followed by more complex malware attribution and correlation queries.

Another key challenge was the wrong motivations for the DevOps team, where MBO was used to evaluate and provide bonuses for staff. This incentivized staff for short-term goals, nothing long-term. They also had a short-term mindset since the development industry was moving towards agile computing with short sprints and away from the long-term planning approach of SDLC. This shift resulted in a generation of programmers that grew up in a culture where they are rewarded for the here and now, nothing long-term, and not many experiences beyond quick sprints. This mindset had to change to include long-term strategic goals and motivations built into how they received financial rewards, bonuses, and growth opportunities. It's not hard, but it is essential, as leaders, to ensure we motivate staff in the right way to incentivize the proper outcomes.

Any time a tech team advertises shiny new capabilities and can't demonstrate actual results in production that is a major red flag! Always trust but verify. Discuss different frameworks, challenges, and approaches with various subject matter experts across the industry.

⊕ Sharpshooter Tip

- Know your people, what motivates them, and how things operate on the floor. Research solutions from all directions – bottoms-up, tops-down, as a team, and unify towards a consistent outcome. This all takes time, but investing in your people with a focus on trust results in positive change, especially when starting in a toxic culture.

⌐⃗ Misfire

- If anyone tells you the latest shiny tool will solve all your problems or do "all these amazing things," your internal red flags should go up. Find trusted colleagues to verify, do your homework, and approach cautiously.

⊇ Data loss watching

I was a CISO at a technology company that held government contracts. After assessing the technology environment, I found that we did not have data loss prevention (DLP) or data classification in place. I worked on a roadmap and budget for both projects and presented the strategy and implementation plan to executive management.

One VP was very supportive and stated, "At least two new technology companies have been founded using our intellectual property (IP), which was taken from our company." Unfortunately, I received pushback from the VP of Engineering, who was concerned about the program slowing down his projects.

Considering the VP of Engineering's concerns, the team worked to put the tool into detection mode for an extended time. This allowed IT to test whether the endpoint agent interfered with other software in the environment. In addition, it allowed the newly established DLP team to get a baseline. Implementation began with a small number of individuals within the Information Security Team and spread to the entire department.

As part of the rollout, I communicated with the department, describing the tool and how it worked. Because the tool was in detection mode, the alerts could be better described as "data loss watching," but by the time we investigated an alert, the data had already moved. We could start to see how information moves within the environment. After 6 months of testing, the head of the DLP unit pulled me aside to discuss a concerning alert. Yesterday, one of the senior engineers moved a terabyte of files onto a personal USB drive.

Sharpshooter perspectives

SSP1: What intellectual property (IP policies are in place? I assume there are policies in place that prohibit taking company documents. If the investigation determines the download contains company files, legal ramifications for the individual may occur. However, without an investigation, I would not presume that the files are company-related. I've seen company-owned computers with enough video files to rival Netflix, so this could be something other than IP theft.

SSP2: "The lady doth protest too much, methinks," as Hamlet might say about the VP of Engineering. Sometimes, people know they are doing something they shouldn't but still do it because they feel a sense of entitlement that specific rules don't apply to them. Developers and intellectual work product teams seem to use this pattern more frequently. When someone spends a lot of their intellectual capital in creating something, a sense of personal ownership remains even when it is obviously the IP of the organization that paid them to do it. I am not saying this is the case here, as I have seen some less malicious things cause similar alerts. However, if it looks like a duck, quacks like a duck, and swims like a duck, it is reasonable to assume it is a duck.

Target grouping

The DLP unit had detected an engineer loading a terabyte of files onto a USB drive. It was discovered that the senior engineer had received a position at a competitor but had not disclosed that they were leaving. The information they had moved to the USB was company-proprietary, and the investigation took several weeks to determine the extent of the files they had taken. They were fired from their current position, and legal action was threatened unless they returned the files.

⊕ Sharpshooter Tips

- Implementing DLP software shifts the cultural shift to protect the company first. Many executives hold an optimistic belief that most employees are ethical and will not take company property.
- During implementation, be prepared for your team to find individuals stealing company information. Work with HR on the disciplinary process. Because much of the theft occurs when employees leave, coordinate with the Legal Team as they may pursue legal action against the former employee.

⌐══ Misfire

- Many DLP implementation plans underestimate the resources needed to review and follow up on alerts. In this case, an internal team and the MSSP handled the workload, avoiding a misfire.

Index

Printed in the United States
by Baker & Taylor Publisher Services